IMAGES
*of America*

# SPRINGFIELD

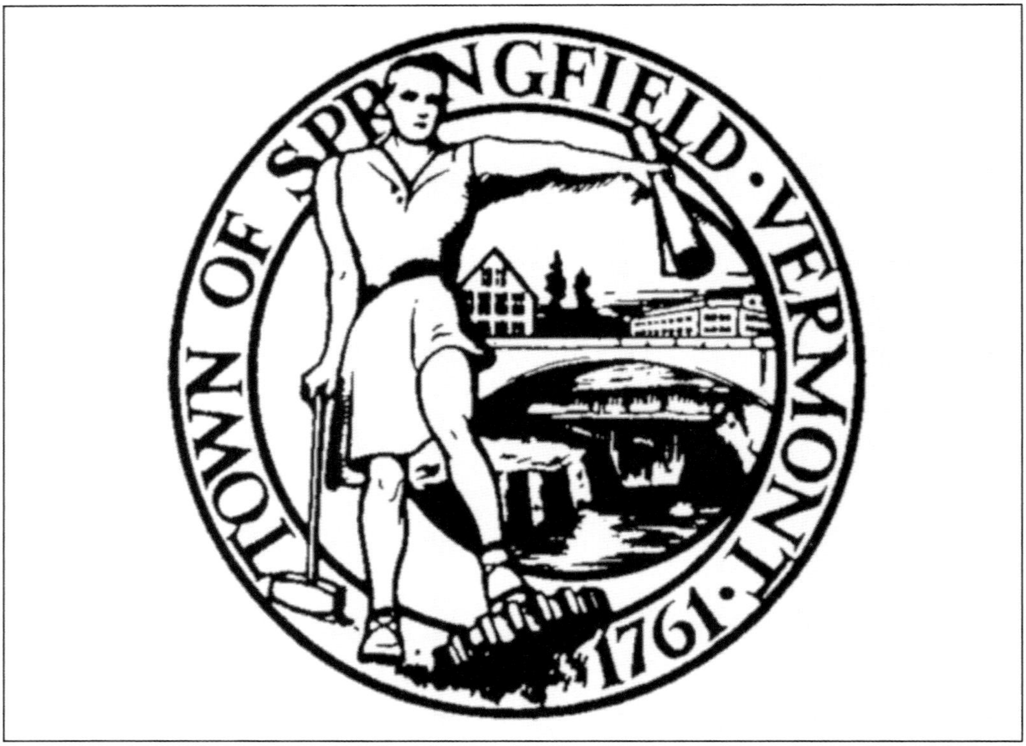


Russell Porter designed the Springfield town seal in 1921. The arched bridge spanning Comtu Falls is the source of power for this industrial town. The youthful male holds a hammer symbolizing labor and a scroll symbolizing invention. He is standing on a gear representing the various manufacturing businesses. The town was chartered in 1761. The pine trees reflect the Vermont state seal. (Springfield Town Office.)

**On the Cover:** This is a Buster Brown Shoe promotion at the C.H. Moore Store that was located on the south end of the Woolson Block. C.H. Moore was in business from 1901 to 1925, but the Buster Brown character lived on and made it to television. Buster's catchphrase was, "Hi, I'm Buster Brown, I live in a shoe. This is my dog Tye, he lives there too." (Springfield Town Library.)

IMAGES
*of America*

# SPRINGFIELD

Rosanne E. Putnam

ARCADIA
PUBLISHING

ISBN 978-0-7385-7518-6

Published by Arcadia Publishing
Charleston, South Carolina

Printed in the United States of America

Library of Congress Control Number: 2010940354

For all general information, please contact Arcadia Publishing:
Telephone 843-853-2070
Fax 843-853-0044
E-mail sales@arcadiapublishing.com
For customer service and orders:
Toll-Free 1-888-313-2665

Visit us on the Internet at www.arcadiapublishing.com

*To my husband, Hugh, who was the true force behind this whole project. It is his love of history and appreciation for how important it is to remember and respect it that sent me down the road leading to this book.*

# Contents

# Acknowledgments

If I could only acknowledge one person, it would be my husband, Hugh. No one did more to help with the writing of this book. Not only is he a wealth of knowledge, but he is also a local history pack rat. If I had a question, he would disappear into the closet and come out with a piece of paper, pamphlet, or book about the subject. I could not have done this without his help.

Most of the photographs in this book came from three sources. The images marked STL came from the extensive collection at the Springfield Town Library, which also includes Fred Richardson's postcard collection. Those marked WGS are from Harland Whitcomb's glass slide collection now housed at the Vermont Historical Society in Barre. Items marked HSP are from our personal collection. Those marked BL are photographs taken by Bernie Lashua, a longtime local photographer who had the good sense to go around and document life in Springfield over several decades. Thank you to Bill Lashua, Bernie's son, for making these photographs available. The Springfield Art and Historical Society provided the photographs marked MAC, and I would like to thank Ken Stringham of the society for his help. There are also a few items from Scott Andrew Bartley that are marked SAB. All other images are individually marked.

So many people have been incredibly helpful. Among them are Warren Garfield, who gave me complete access to the Lovejoy Tool historical information; Kaye Peck, who was very helpful with information about and photographs of North Springfield; and Bert Willard from the Springfield Telescope Makers, who lent a hand with the Russell Porter–Stellafane chapter.

No amount of thanks is enough for Pat Graves, who agreed to be my proofreader. Her teaching background and her interest in Springfield history were a winning combination.

# Introduction

According to tradition, in 1752, John Nott and his wife, Ruth, who was believed to be part Indian, came from Connecticut and settled first in French Meadow with the Abenaki Indians. They later moved to the mouth of the Black River, which would locate them across the Connecticut River from the Fort at No. 4, the most northern fort on the river at that time. Since no one owned this land, the Notts cleared a lot and built their home.

In 1759, the British, continuing their perpetual battle with the French in the north around Lake Champlain, decided to build a road across the grants from the Fort at No. 4 to the Crown Point Fort on the western side of Lake Champlain. John Nott, being an industrious man, realized that they would need a ferry to cross the river, and he obliged. The project was completed in 1760.

By now, Gov. Benning Wentworth of New Hampshire was very busy selling off the grants, and in 1761 at a meeting in Northampton, Massachusetts, the town of Springfield was chartered by a group of men who were primarily from Massachusetts and Connecticut. John Nott, one of many squatters, found himself ejected from his homestead.

There were also land disputes as a result of New York claiming it owned all the territory east to the Connecticut River and selling the same grants. King George III ruled in favor of New York in 1764, causing the New Hampshire landowners to lose their land.

Because of the Crown Point Road and the fact that the land by the Black River was a virtual swamp, early settlement of Springfield was in the Eureka area and on Parker Hill. As time passed, however, it became evident that the river and the great falls were where the future lay. Men like William Lockwood and Isaac Fisher could appreciate the prospects along the river. They would fell the huge trees, drain the swamps, and harness the waterpower over multiple dams to turn the wheels of industry in the valley.

To facilitate development of the area, Isaac Fisher built a covered bridge over the Connecticut River in 1806. This was an improvement, but people were still required to travel over Skitchewaug Mountain on the Crown Point Road. In 1812, he built a thoroughfare up the Black River Valley, cutting the trip to about seven miles.

As people moved down from the farms on the hills in the early 1800s, sawmills, gristmills, and blacksmith shops were built. Stores, small businesses, and a hotel were opened. The first two churches on Main Street (Congregational and Universalist) were built in 1833. Although children were being schooled, it would be some time before districts and schoolhouses were established.

There were many settlers in Springfield who were not just dirt farmers but problem solvers and inventors. Many would receive patents for their inventions. While these items might seem mundane or archaic now, they were life altering at the time. Some of these inventions were the combination lock, spring clothespin, blanket hook and eye, a corn planter, a screw head and driver, a joint for a carpenter's rule, a broom holder, the Taylor mop, a mop wringer, an improved method for polishing marble, a process of weaving cane into webs for chair work, the Warren plow, and sandpaper. The first toy carts, doll carriages, and jointed dolls were also made here.

This list does not include all of the remarkable advancements made in the shoddy (cloth made from reclaimed wool) and cloth processing and finishing businesses.

At the end of the 19th century, a decision was made that would change the landscape of the town for the next 80 years. By 1880, the population of the town was shrinking, as it was all over New England. Jobs were needed to keep people in the area. In order to make Springfield more attractive to workers, the leaders of the town decided to buy J&L Machine Company in Windsor, Vermont, and move it to Springfield. A year later, 27-year-old James Hartness was hired as plant superintendent. These two actions would turn Springfield from a textile-mill town to a world-renowned machine tool industry area known as the Precision Valley. This valley was so well known in World War II that it was among the top 10 targets of the Nazis.

Photography did not come into its own until the mid-1800s, so the earliest photographs in this book do not start until around the time of the Civil War. Much had already happened by then, and every effort has been made to convey that previous history as well as what came after.

# *One*

# Roads, Bridges, and the Railroad

It is necessary to be able to get important assets to any successful town. In Springfield's case, this was accomplished by river. The river's spectacular falls had a 110-foot drop. Its tumbling waters ran through the town for 10 miles. Plying the Connecticut River by ferry to bring supplies was not convenient or efficient, and carrying them up the Crown Point Road, over Skitchewaug Mountain, and down into town compounded the hardship. What was needed was a road up the valley. Once that was completed in 1840, thoughts turned to the railroad. Rail transportation would move goods even faster and easier.

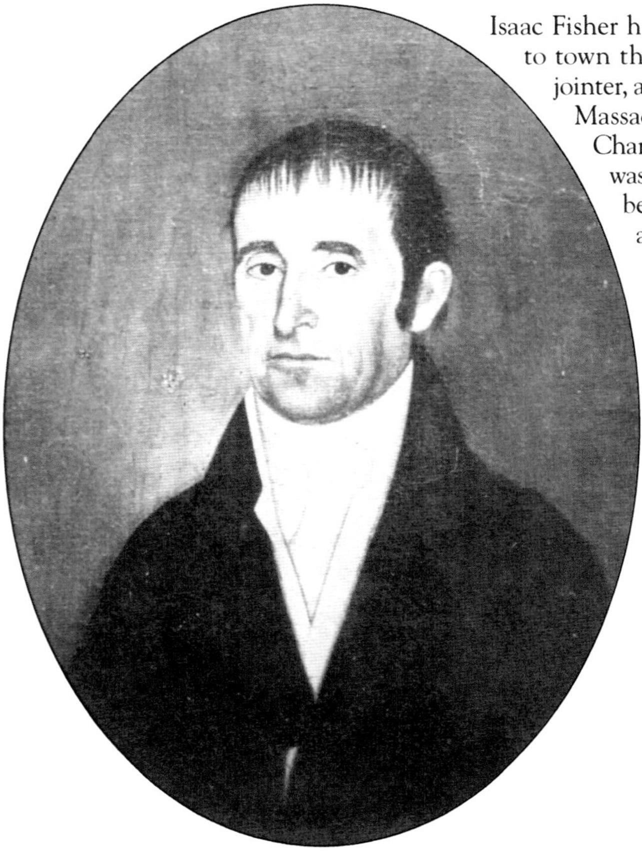

Isaac Fisher had more to do with getting people to town than any other citizen. A carpenter, jointer, and miller, Fisher was born in Natick, Massachusetts, in 1763, and he moved to Charlestown, New Hampshire, in 1799. He was responsible for the first covered bridge between Charlestown and Springfield and the construction of a road up the Black River Valley. (WGS.)

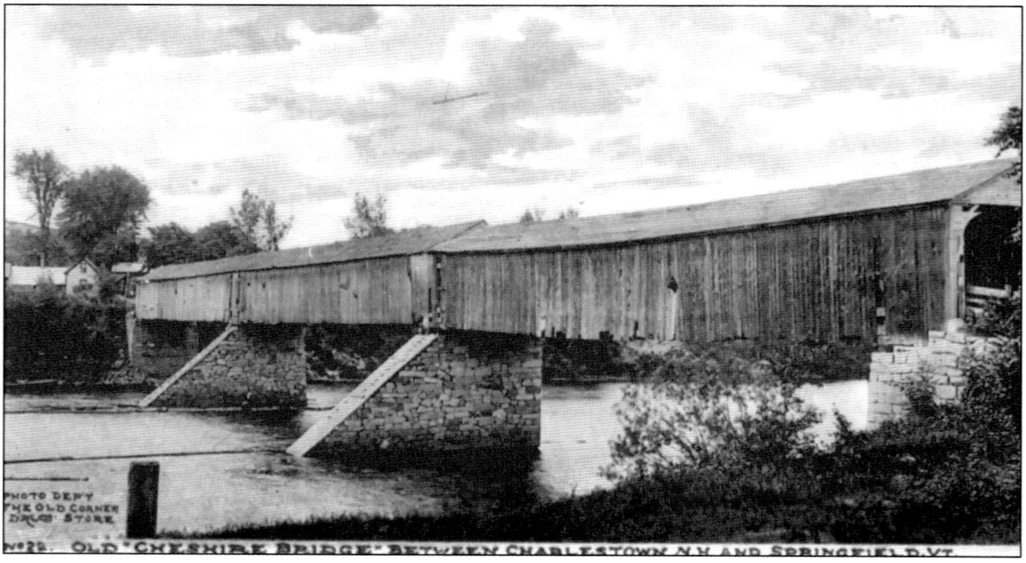


The town of Springfield was incorporated in 1761. At that time, the only way to cross the river from the Fort at No. 4, located at Charlestown, New Hampshire, was by ferry. The Cheshire Bridge Company was formed, and in 1806, under the direction of Isaac Fisher, the first covered bridge was built across the river. Subject to many ice jams, the two left sections had to be replaced in 1862. (STL.)

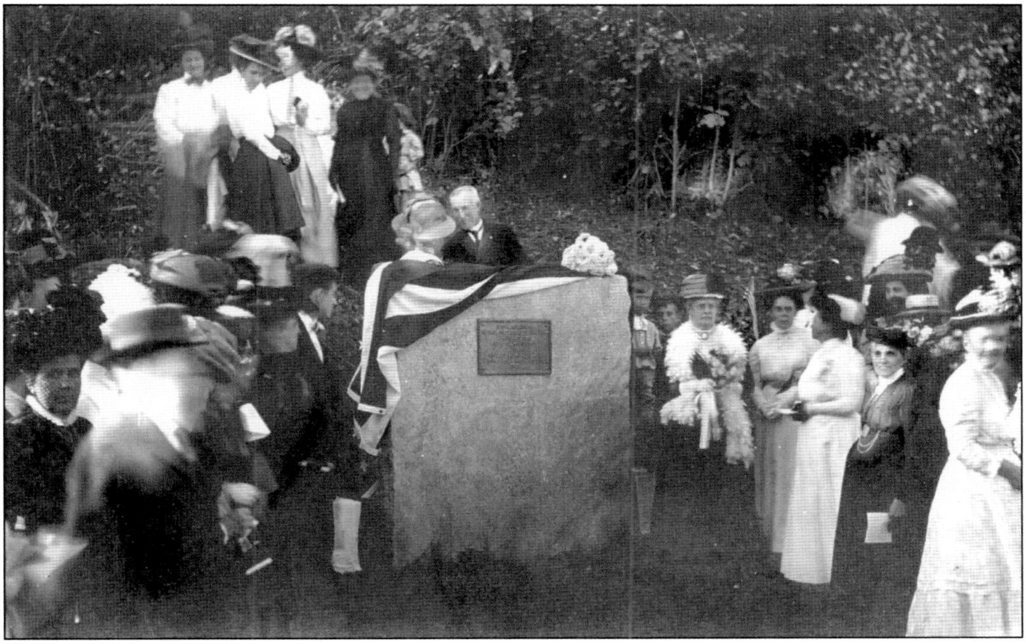

After crossing the bridge over the Connecticut River, one had to go over the Crown Point Road and Skitchewaug Mountain to get to the town, which was now moving from the hillside down into the valley. It was a long, tedious trip. The Daughters of the American Revolution (DAR) dedicate the marker commemorating the ferry and the original blockhouse at the beginning of the Crown Point Road in this photograph from 1909. (STL.)

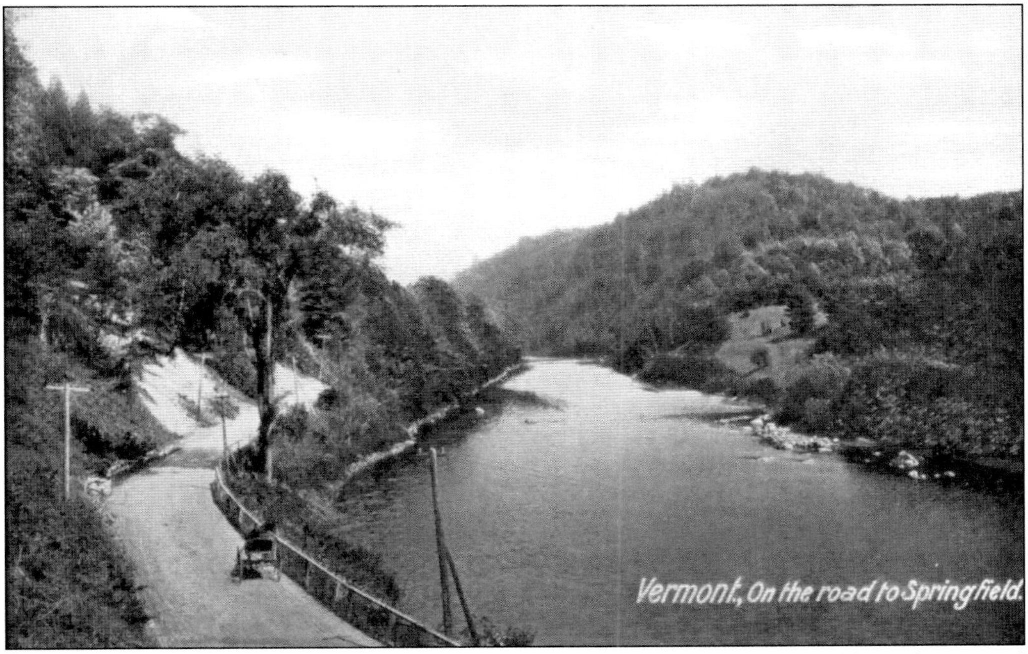


Isaac Fisher moved to Springfield and bought a large amount of land. He could see that a road up the Black River Valley would be a better route than over the mountain. It took many years of volunteer work by local citizens, but in 1812, the road was opened. Not only was it a better route to town, it also opened up the abundant waterpower of the river. (STL.)

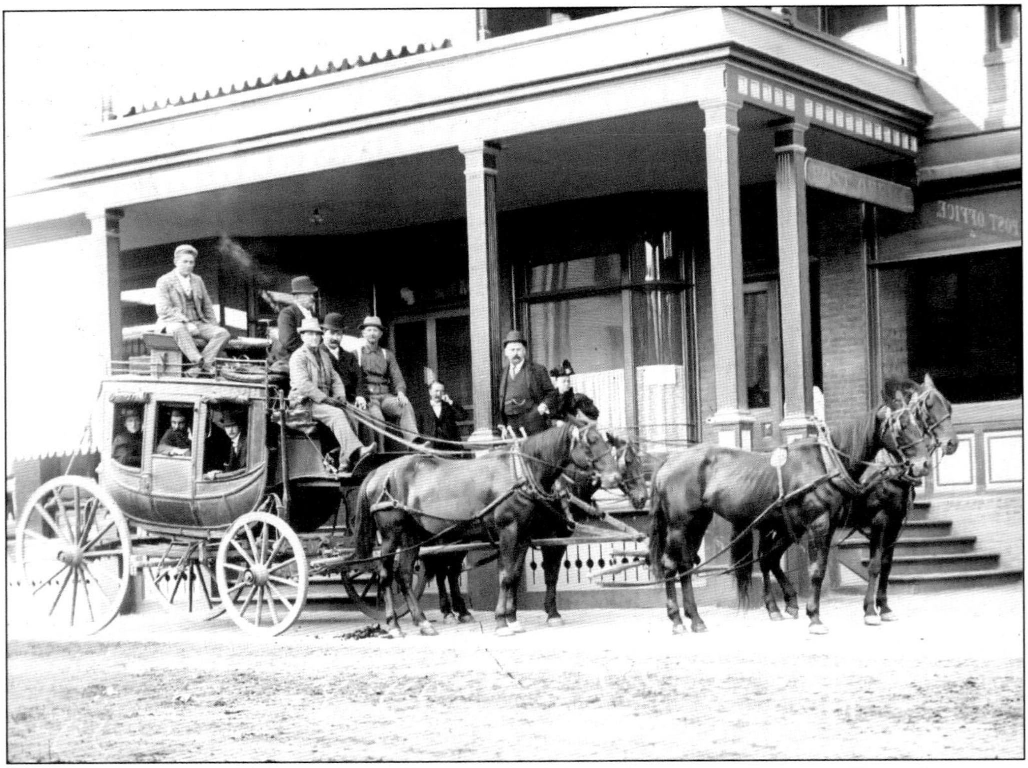

The new road would facilitate carrying passengers from the Boston & Maine Train Station in Charlestown, New Hampshire, to the newly opened Adnabrown Hotel (seen here) in Springfield. Driver Wallace Fairbanks was photographed with Ed Keyes, Sim Long, and John Hart. (WGS.)

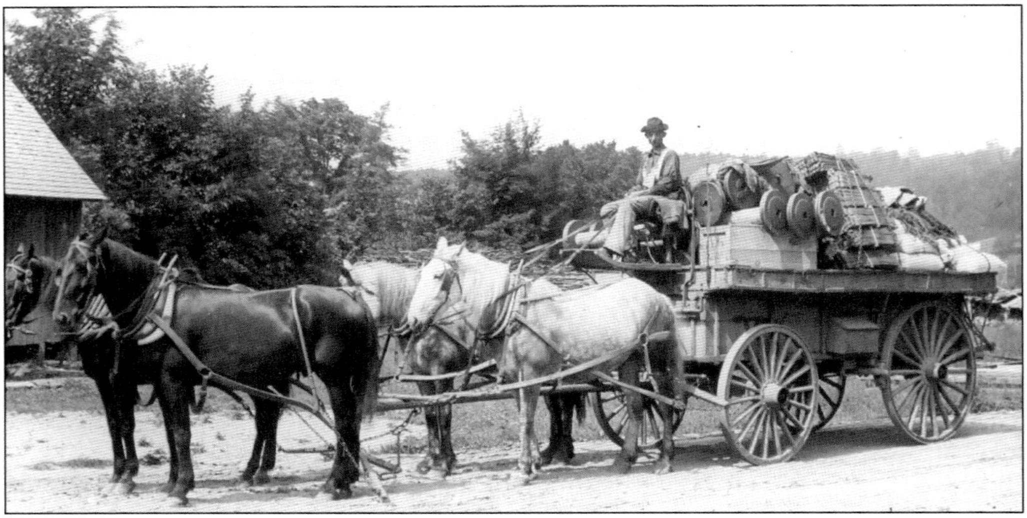

Even with the road, it still was not convenient to carry large amounts of supplies from the railroad station in Charlestown. Henry Parker is the driver of this wagon loaded down with supplies. (WGS.)

The railway was near Springfield as early as 1840, but it wasn't until 1896 that the town's railroad charter was granted. It took a lot of men, horses, wagons, and hard labor to widen the road up the valley to make room for the railroad tracks. In July 1897, Springfield's first passenger service was offered, and in August, the stagecoach made its last trip. (STL.)

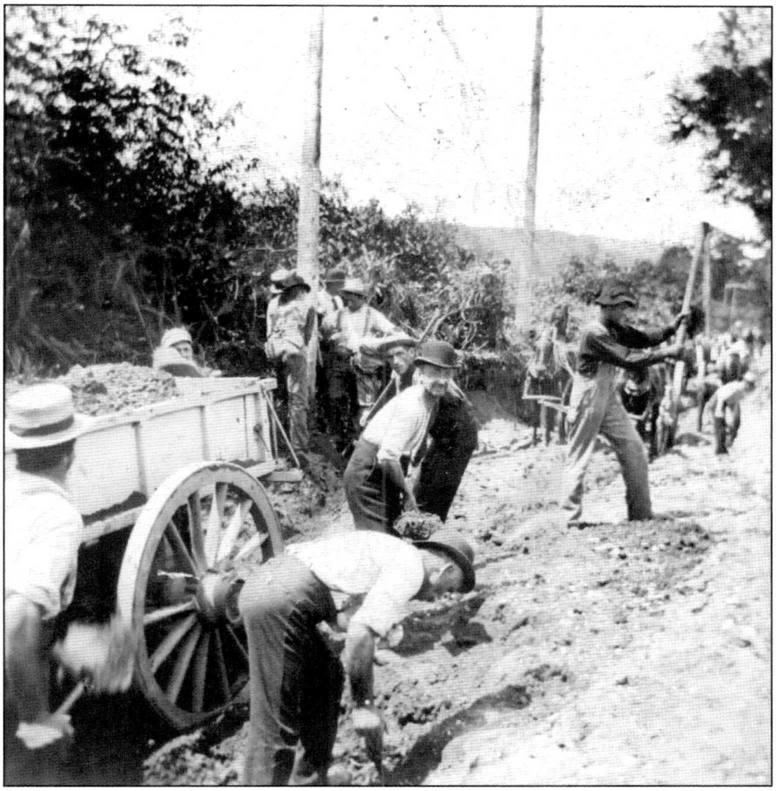

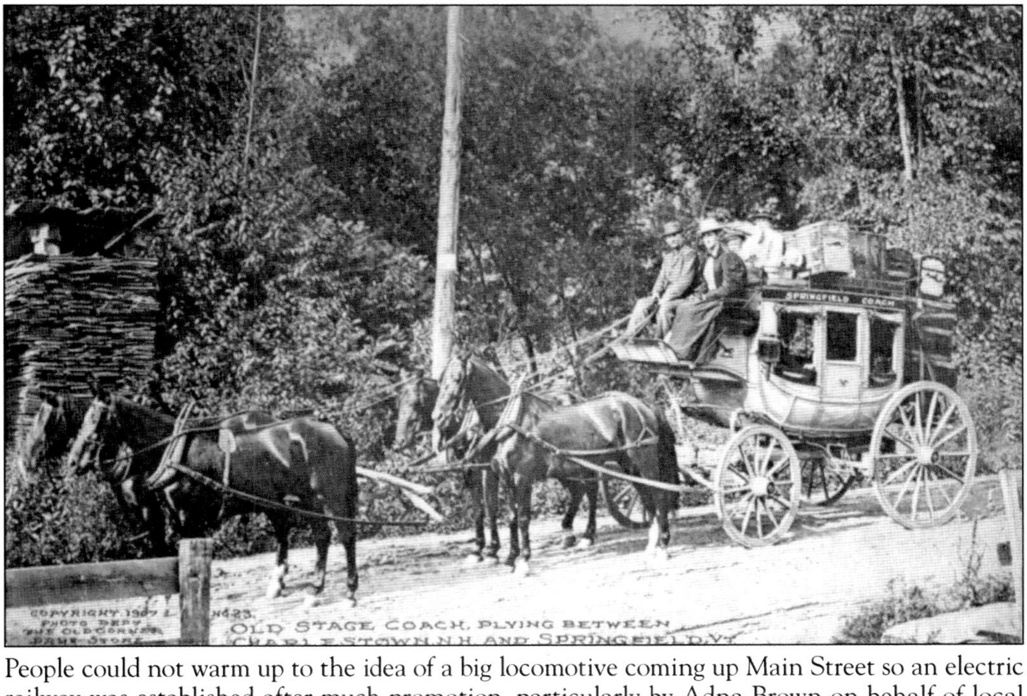


People could not warm up to the idea of a big locomotive coming up Main Street so an electric railway was established after much promotion, particularly by Adna Brown on behalf of local businesses and his hotel. This would mean the end of the long, dusty, bumpy, sometimes hot and sometimes cold ride on the Concord Coach. (HSP.)

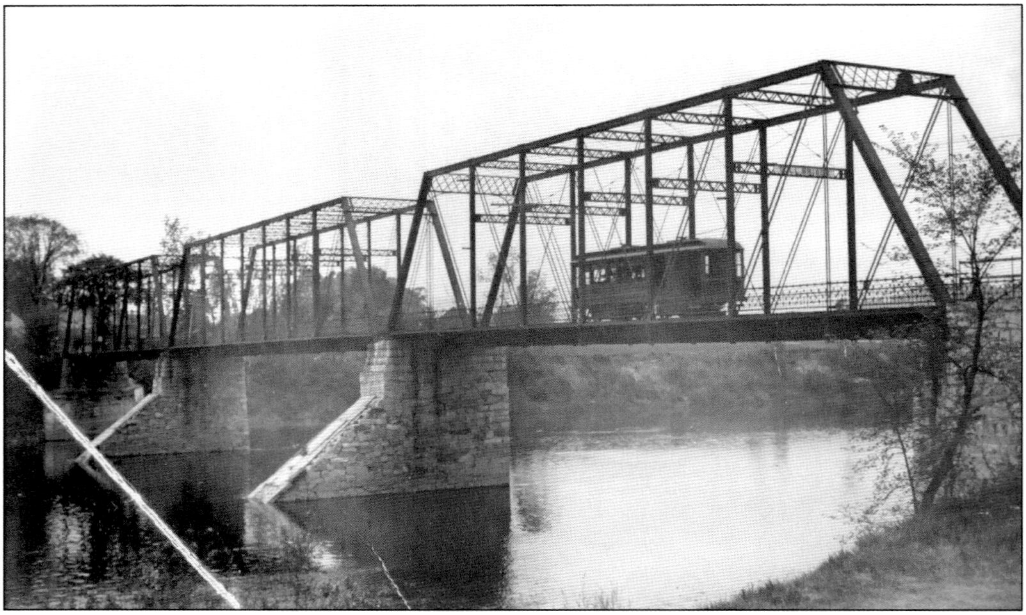

Because the trolley was far heavier than a horse and wagon, it was necessary to replace the covered bridge with a stronger, wider, iron bridge. Built in 1896, this bridge would suffice until 1930, at which point an even sturdier one was needed. (STL.)

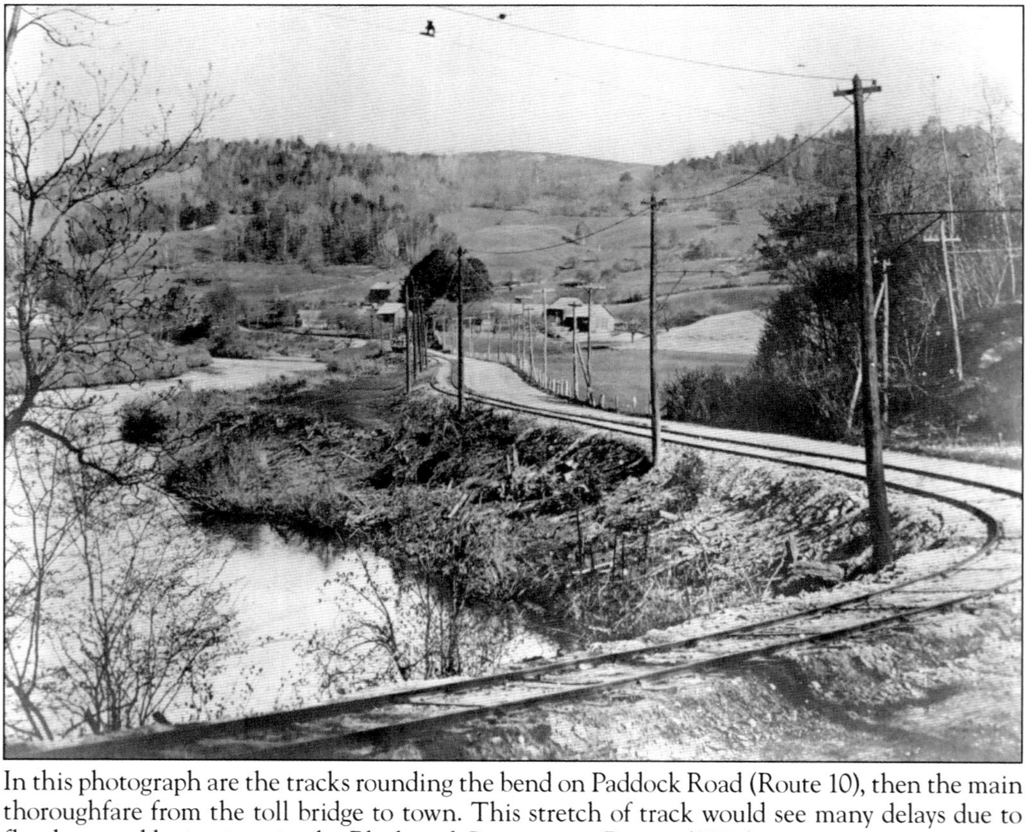

In this photograph are the tracks rounding the bend on Paddock Road (Route 10), then the main thoroughfare from the toll bridge to town. This stretch of track would see many delays due to floods caused by ice jams in the Black and Connecticut Rivers. (STL.)

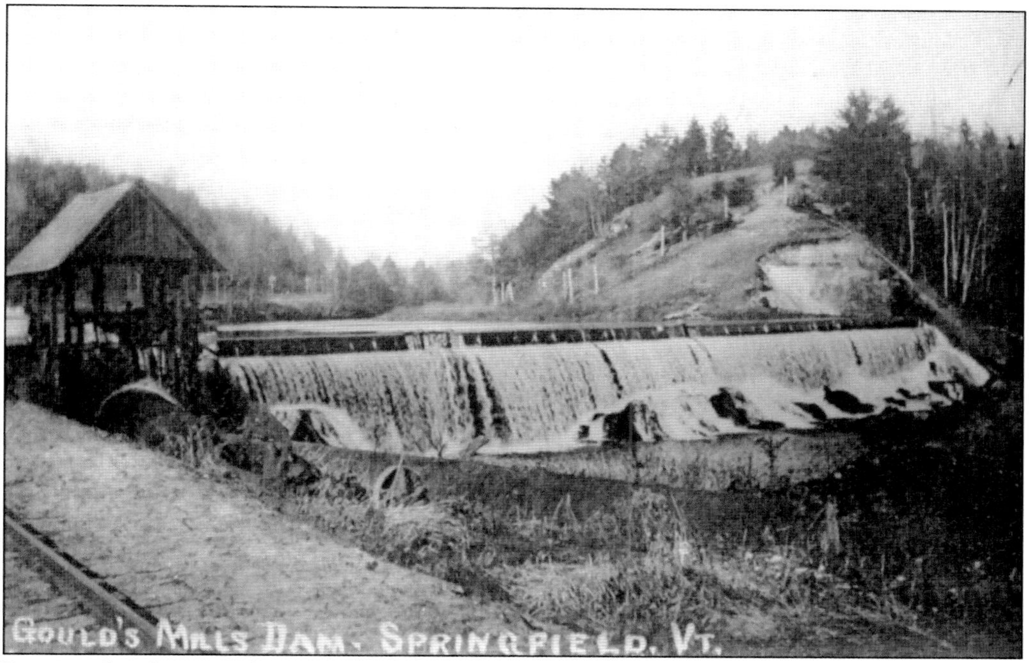


Because the electric railway would require more electricity, the former mill dam at Goulds Mills was converted to produce electricity. (HSP.)

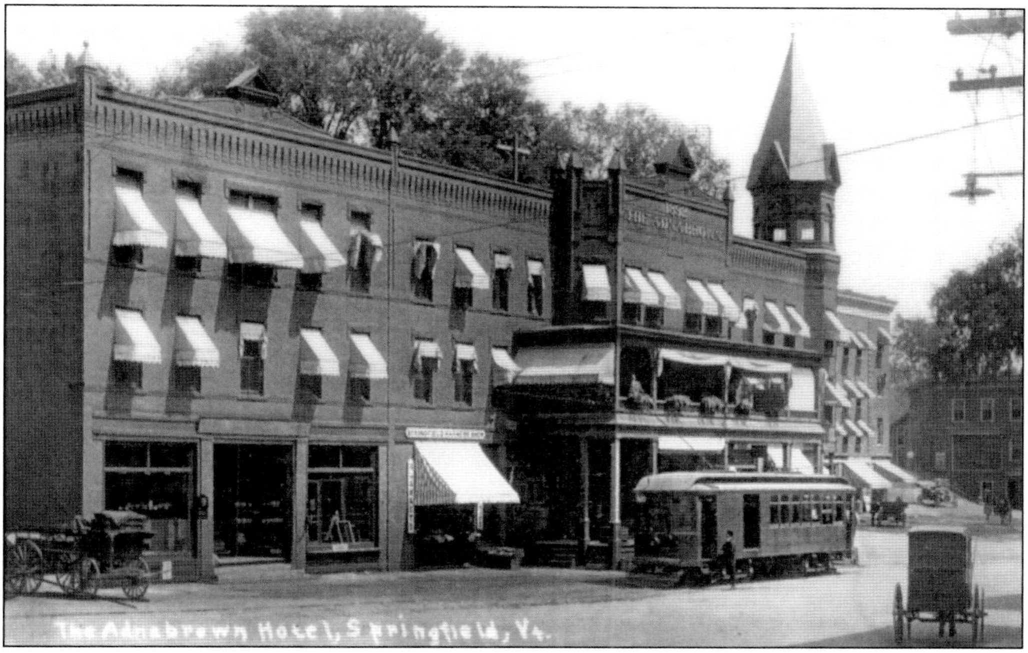


Passengers could now get on the trolley at the hotel or various stops along the way (indicated by white paint on the bottom of telephone poles) and have a smooth, comfortable ride to their destination, whether it was the train station in Charlestown or somewhere else along the way. Notice that in addition to the wagons, there is an automobile coming down Summer Hill at far right. (STL.)

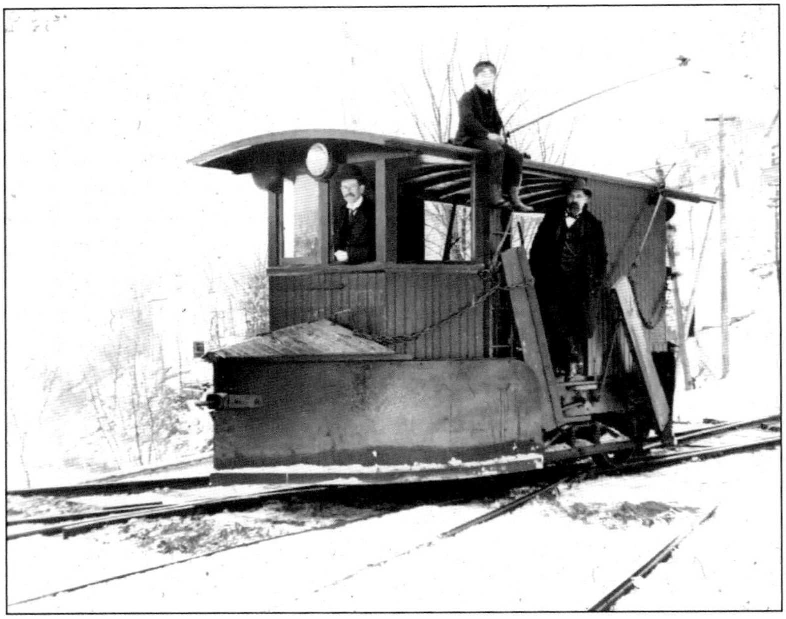

As everyone used a sleigh in the winter, there had been no real need to keep the streets cleared of snow. However, it became necessary to plow the tracks for the train. One would hope the young man on top of the track plow was posing and did not sit there next to the electrical connection when the train was running. (WGS.)

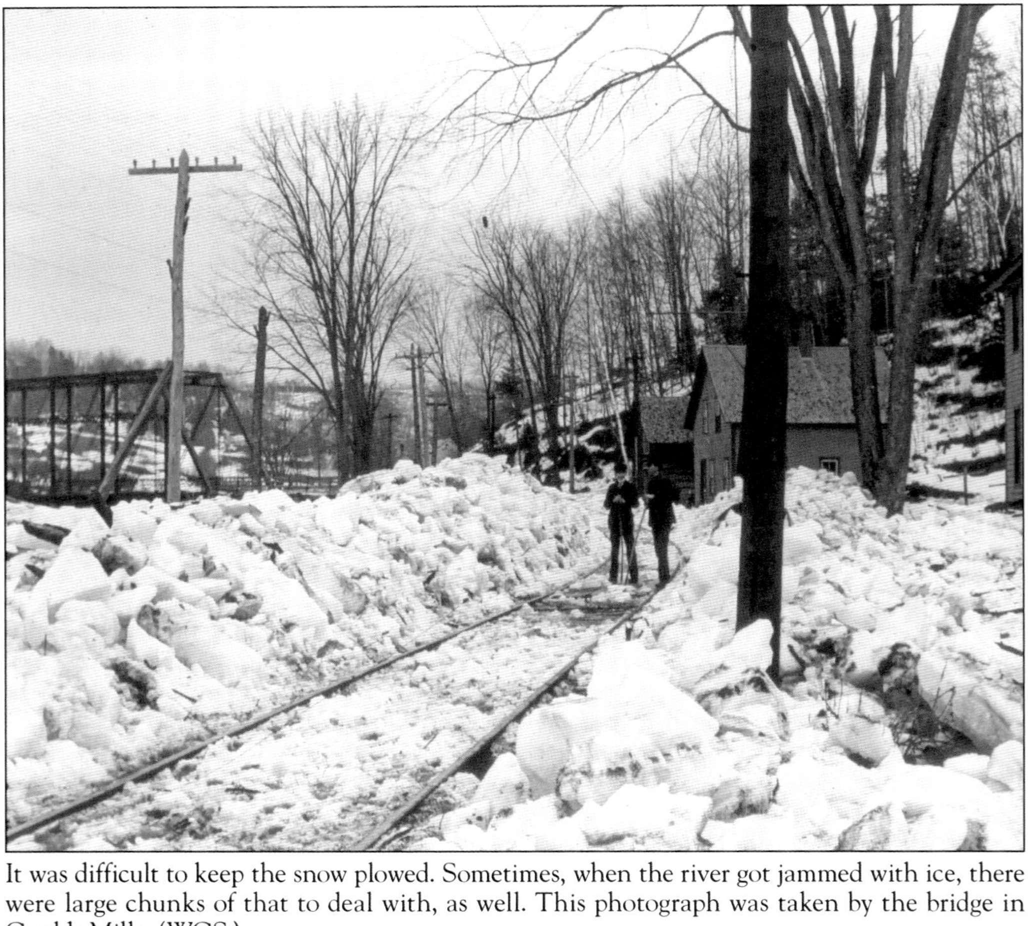

It was difficult to keep the snow plowed. Sometimes, when the river got jammed with ice, there were large chunks of that to deal with, as well. This photograph was taken by the bridge in Goulds Mills. (WGS.)

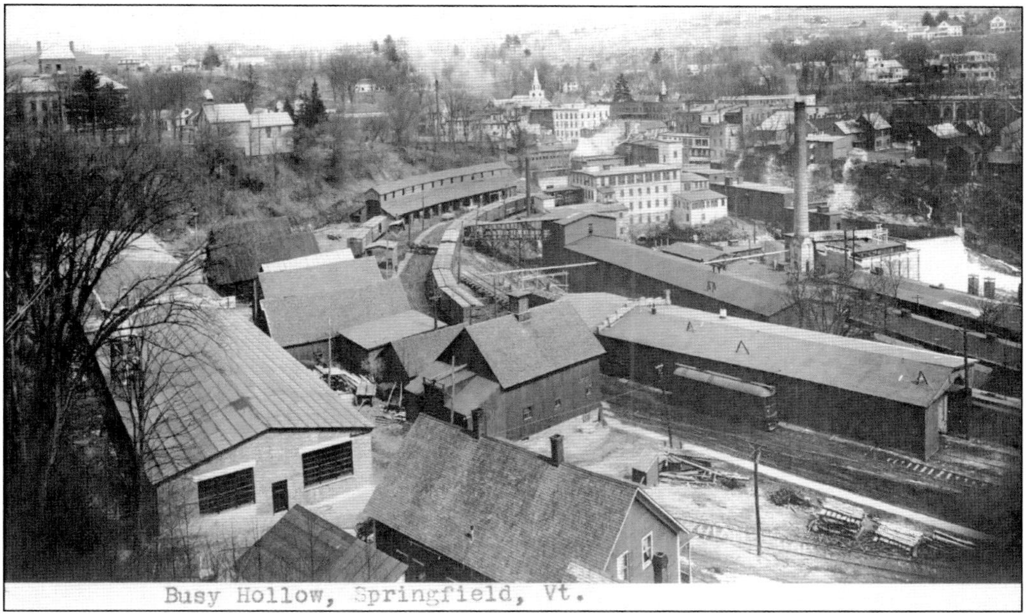


Between the shoddy mill, the coal yard, and the trolley and freight yards, Mineral Street could have been called Busy Hollow. This photograph can be dated to before 1916 because the Methodist Church only has one steeple. The long building on the left is now part of the state office buildings. (STL.)

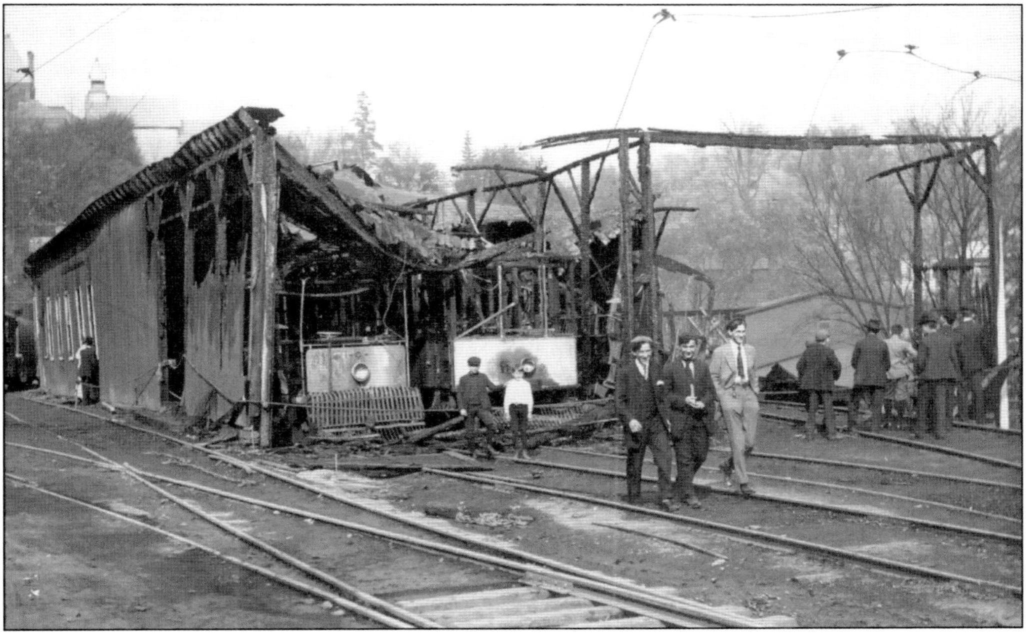

In October 1908, a fire at the carbarn destroyed all five passenger cars, sparing only two freight motors that were parked outside. The alarm whistle was delayed because the electric light company, generating power with a waterwheel, did not have enough steam to blow it. It was January 1909 before the new No. 10 was delivered from Springfield, Massachusetts. Until this delivery, they used the freight engines to transport people. For the midnight run, they used an automobile. (STL.)

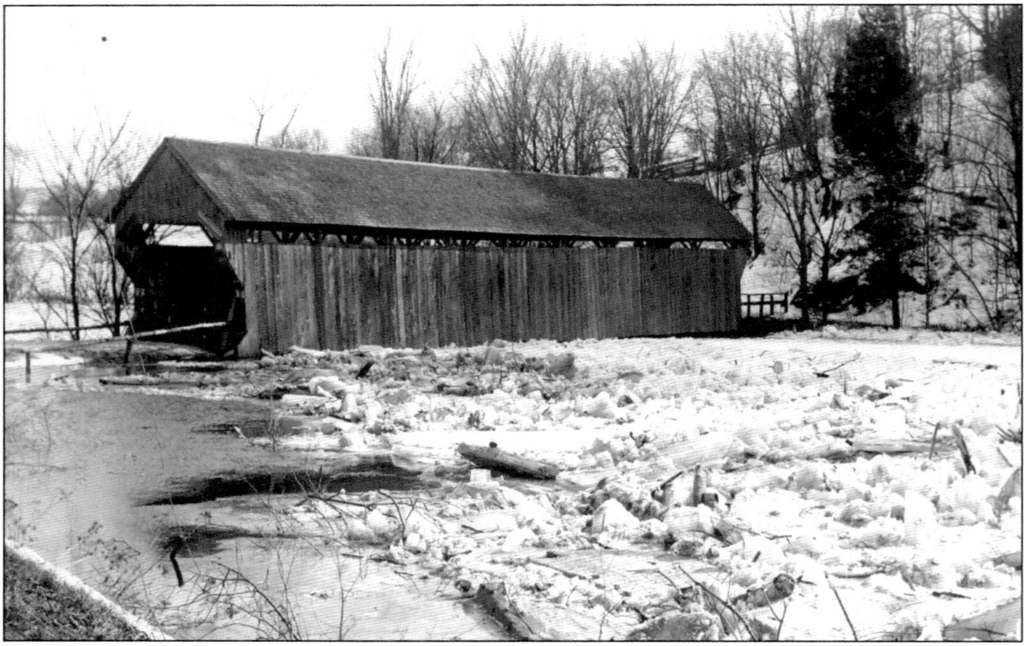

Along with the toll bridge, there were three other covered bridges coming into town. The Goulds Mills and Clinton Street Bridges had to be reinforced and ultimately replaced because of the trolley. This bridge, on the other hand, lasted until 1913. It crossed Black River from the south near what was then the White farm. It was the forerunner to the Missing Link Bridge on Route 5 and faced its share of high water and ice jams. Before the new construction on Route 10 into Springfield, one could see the old foundation. (Above, STL; below, WGS.)

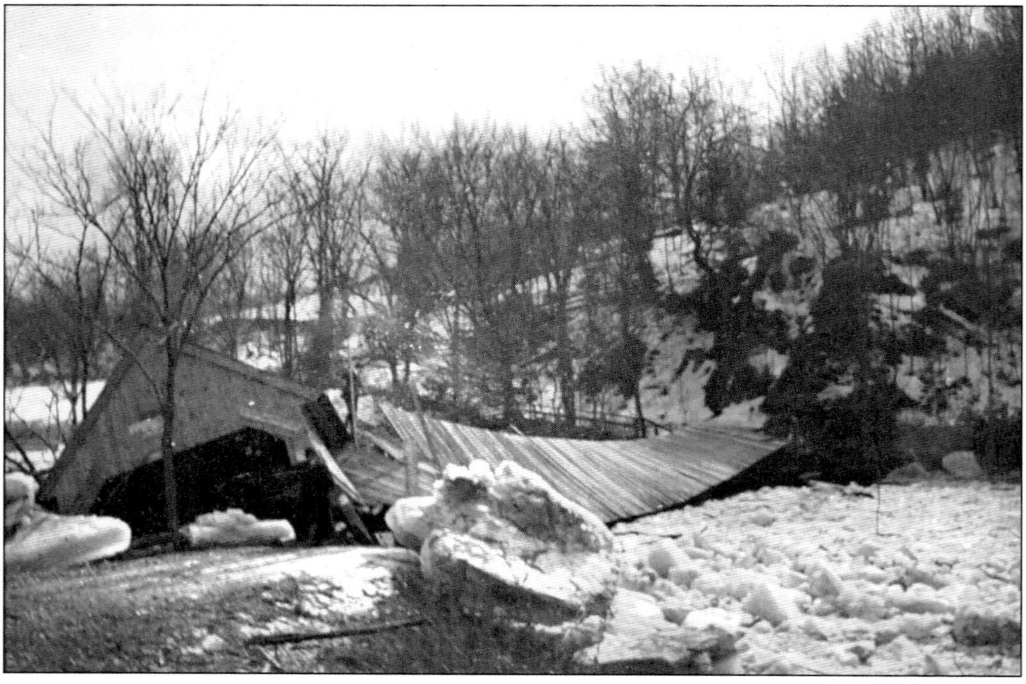

# *Two*

# MAIN STREET

The earliest homes and the first Main Street were on the hill in the Eureka area of town. The land by the river was swampy and, by most people's estimates, uninhabitable. But the early settlers realized that their future lay in the power of the river and slowly moved down the hills, drained the swamps, and built a town.

Main Street was a collection of dry goods stores and other shops, saw and gristmills, a hotel, a couple of churches, and offices, including those of lawyers, doctors, druggists, and craftsmen. In the late 1800s and early 1900s, larger wood and brick buildings began to replace individual homes and shops. In 1895, the library was built. Until then, the books were housed by the librarian—usually a prominent man in the community.

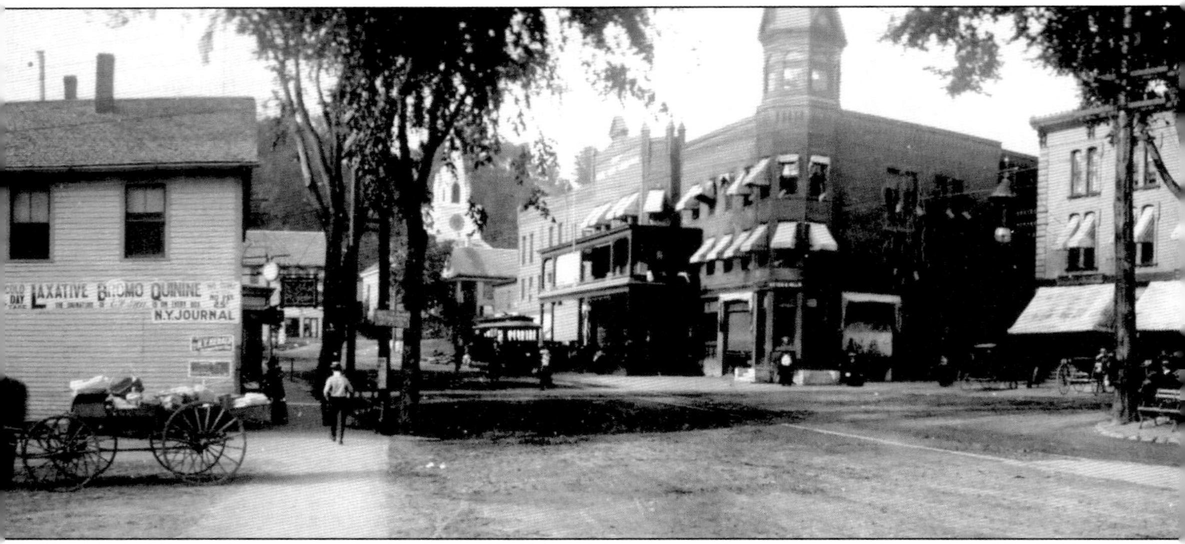


This wonderful view of the square was taken after the trolley but before the remodeling and renaming of the Lincoln/McKinley Block, sometime between 1897 and 1903. On the left, a sign on the Wheeler Building guarantees that Laxative Bromo Quinine will "cure a cold in one day." Across the street, the Boston Store advertises millinery. The house to the right with eight

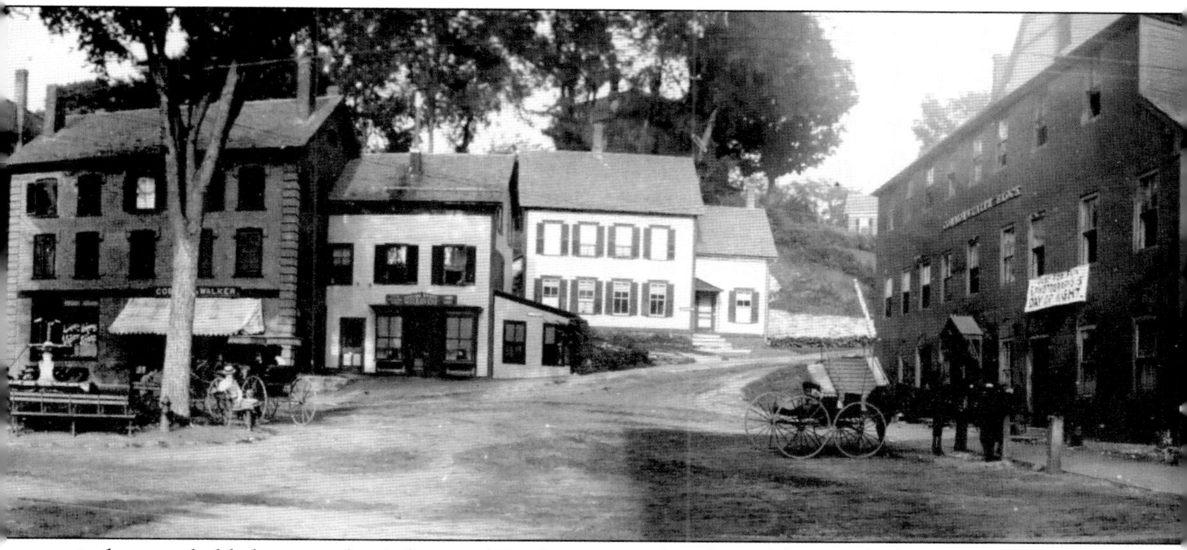


windows probably became the Arlington Hotel mentioned in the *1905 Springfield Directory*. To the left of the millinery is Colburn and Walker, located in the Chase Block. To the far right in this photograph, a sign on the Commonwealth Block reads, "Just for fun photographs, day or night." The movie theater Dreamland would not open in this building until 1909. (STL.)

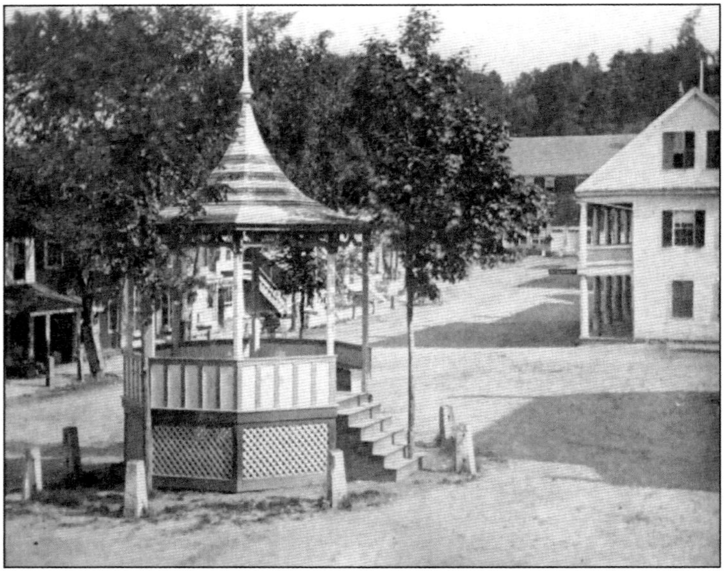

This structure was originally located by a spring on Mineral Street. It was built for several prominent ladies who thought the waters of the spring were beneficial to their health. A group of young men borrowed a four-horse team and brought it to the center of the square as a Fourth of July prank. Much to their amazement, the people liked it, modified it into a bandstand, and left it there until approximately 1879. (WGS.)

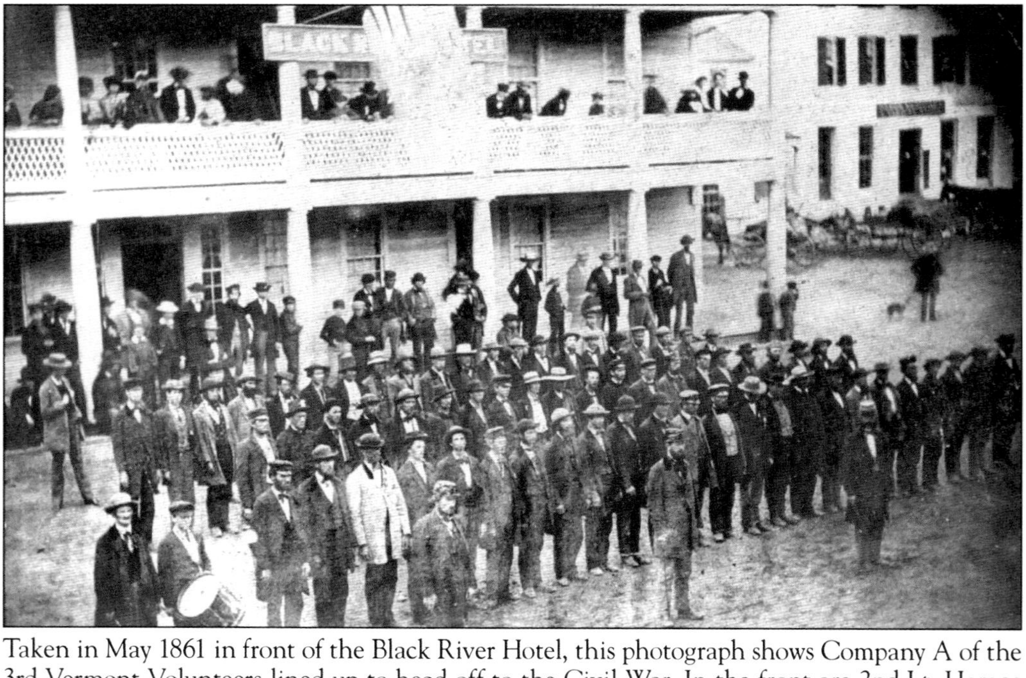

Taken in May 1861 in front of the Black River Hotel, this photograph shows Company A of the 3rd Vermont Volunteers lined up to head off to the Civil War. In the front are 2nd Lt. Horace W. Floyd, 2nd Lt. Frederick Crain, and Capt. Wheelock Veazey. At the extreme left is Seymour Cook (with the fife), and next to him is Horace Bishop (holding the drum). (WGS.)

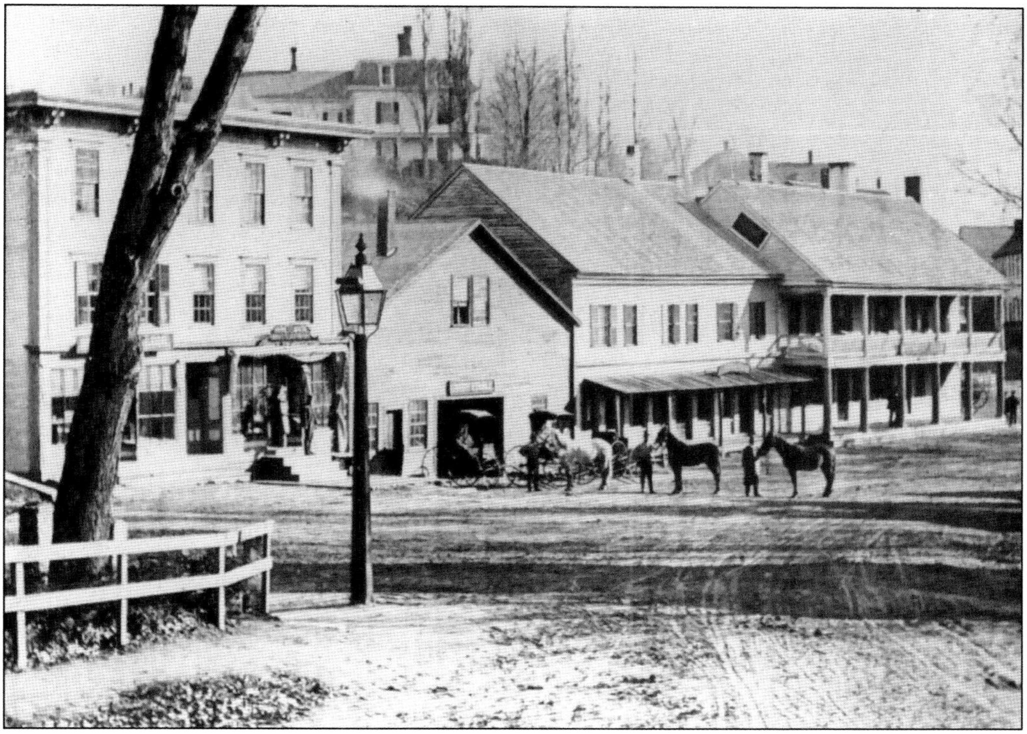

Known as the Springfield House and close to 80 years old, it was time for a modern hotel. The old hotel and livery stable in this photograph were torn down. Adna Brown had the three-story business block moved across the street, where it would become known as the Brown Block, the Lovell Block, or the Bakery Building. (WGS.)

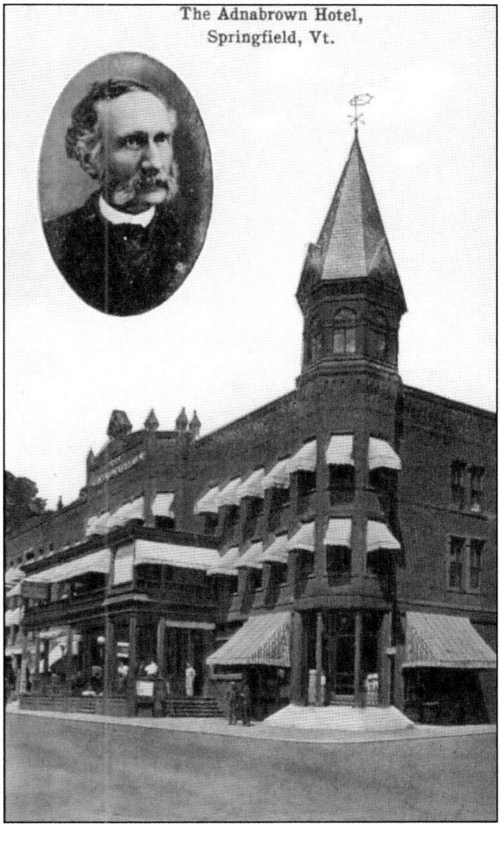


The Adnabrown Hotel was the centerpiece of the square for 69 years. It was built in 1892 to replace the worn-out Springfield House, which was an updated Black River Hotel. It was billed as state-of-the-art because it had electricity, steam heat, and electric bells. Over the years and through incarnations, the hotel would serve patrons of the Concord Coach, the electric trolley, and finally, those arriving by automobile. (STL.)

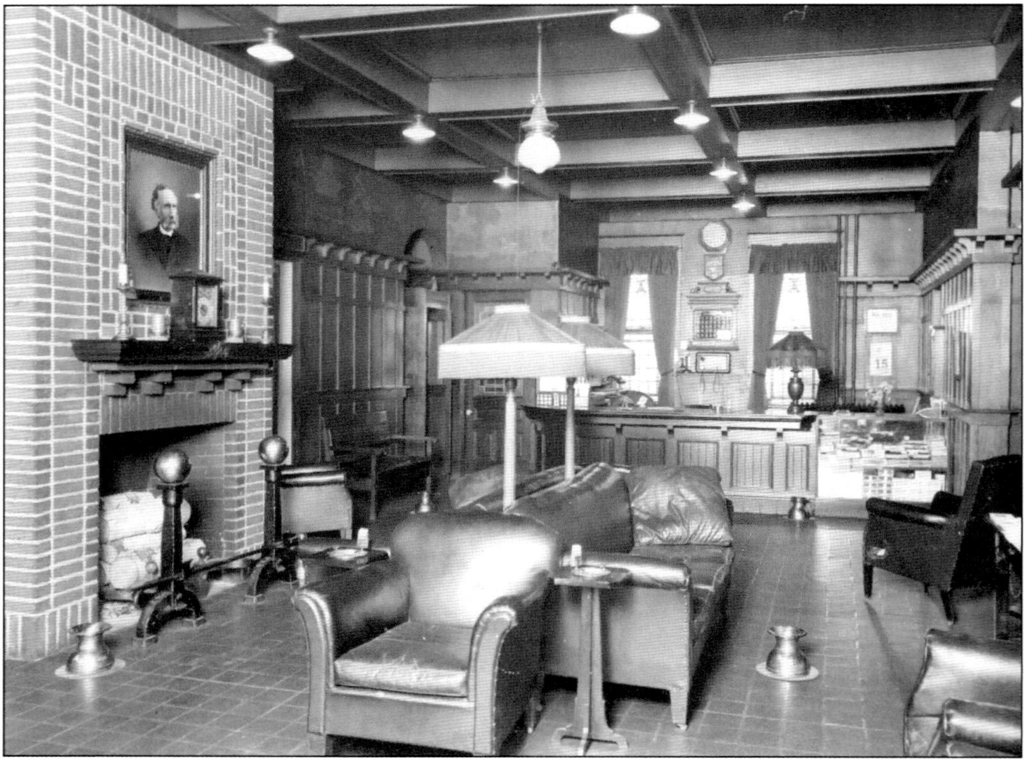

Adna Brown was never a shy man, and his portrait hung over the fireplace in his hotel. The leather chairs, brass spittoons, and stained glass windows are decidedly Victorian. In 1955, the hotel almost was destroyed when the staff decided to burn the Christmas tree in this fireplace and accidentally started a chimney fire. This fire was a precursor to what would happen six years later. (STL.)

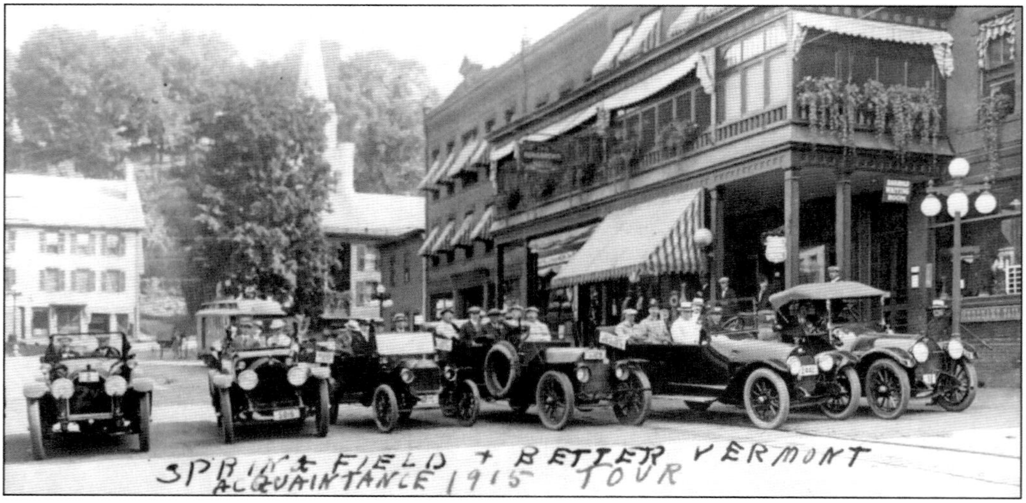


With the automobile came touring. This was an exciting hobby for those who could afford it. The roads were dirt or mud, and gas stations were not nearly as prolific as they are today. During the summer of 1909, the Adnabrown Hotel estimated that it had entertained more than 217 automobile parties. This photograph of mainly Springfield men and their "machines," as they were called, was taken in 1915. (STL.)

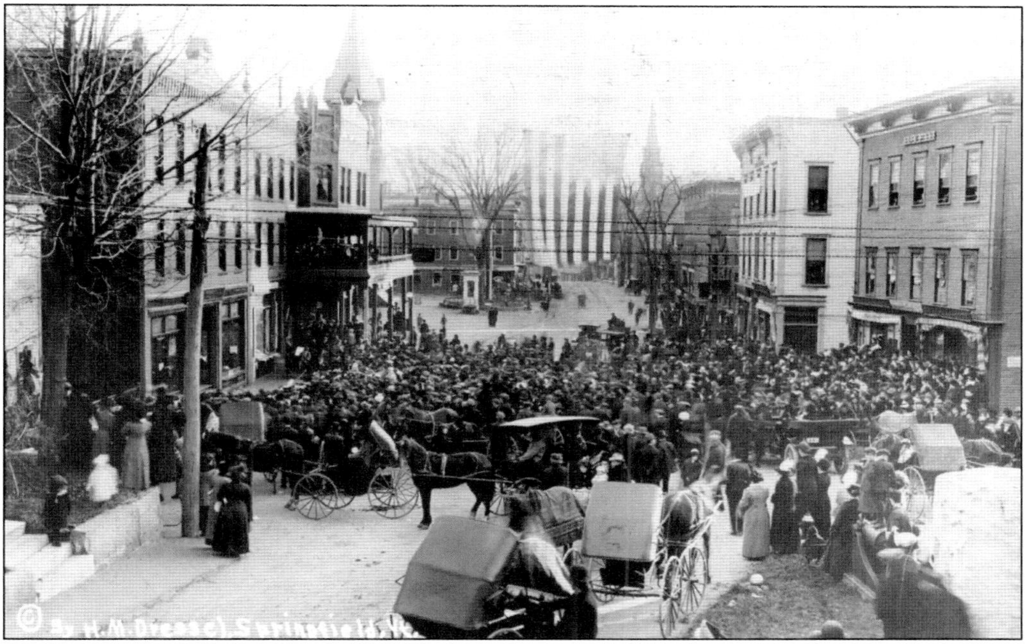

On December 6, 1913, the merchants decided to kick off what was called the Third Annual Opening Day of the Christmas Season with a wedding in the double windows of the LaFountain, Woolson & Co. Hardware Store located in the hotel building. The service went smoothly, according to the *Springfield Reporter*, "without any awkwardness and break in the arrangements." Local merchants donated many gifts, and the whole thing was considered a success. (HSP.)

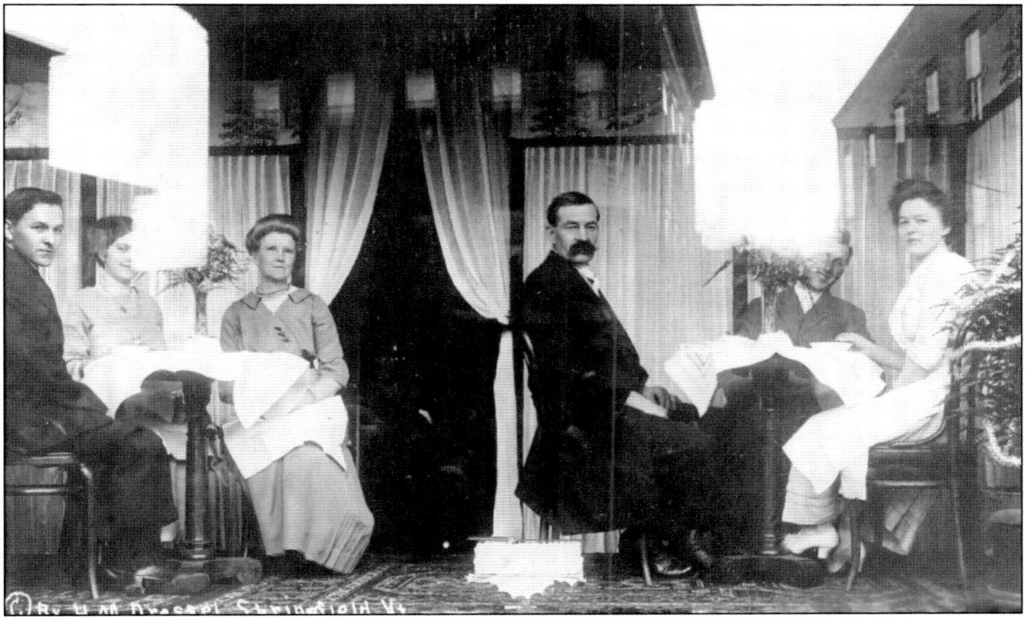

The ceremony for Luther Hackett and Lulu Taft was performed in the left window, and the wedding luncheon took place in the right window. After the luncheon, the couple left in the Woolson limousine. The wedding drew a very large crowd, but it also drew some very disparaging remarks from the *Brattleboro Reformer* newspaper, which objected to using a wedding ceremony as a marketing gimmick. (HSP.)

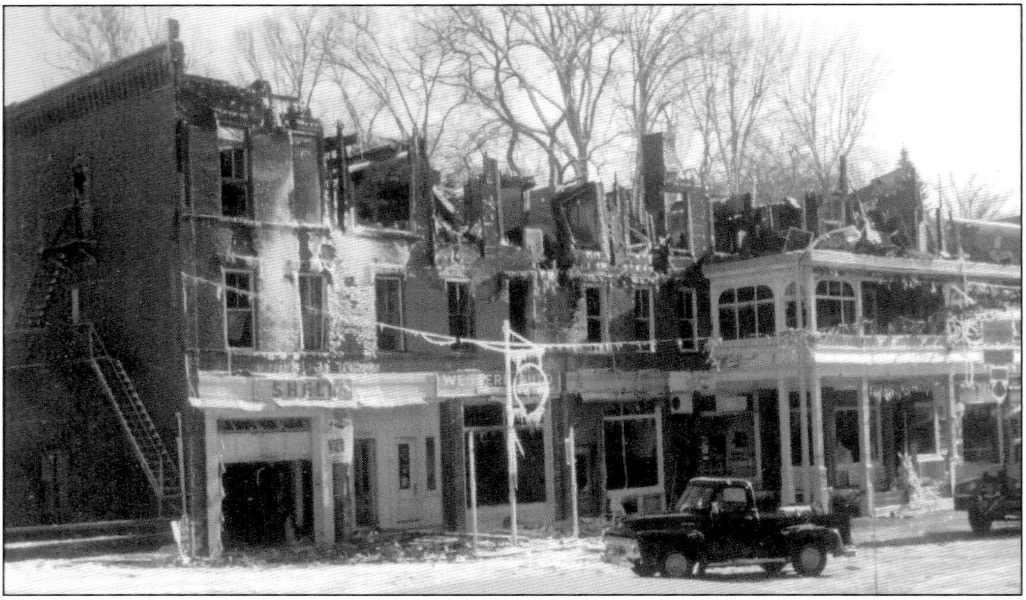

The Adnabrown burned down in 1961 on New Year's Day. It was booked to capacity with 117 guests, most of whom were skiers. Fortunately, it was 10:30 a.m., and most everyone was out of the building when the fire broke out. The flames started in a trash bin in back of the building and went up through the walls and into the roof. It was a loss for the hotel and for the town; the site is now a parking lot. (HSP.)

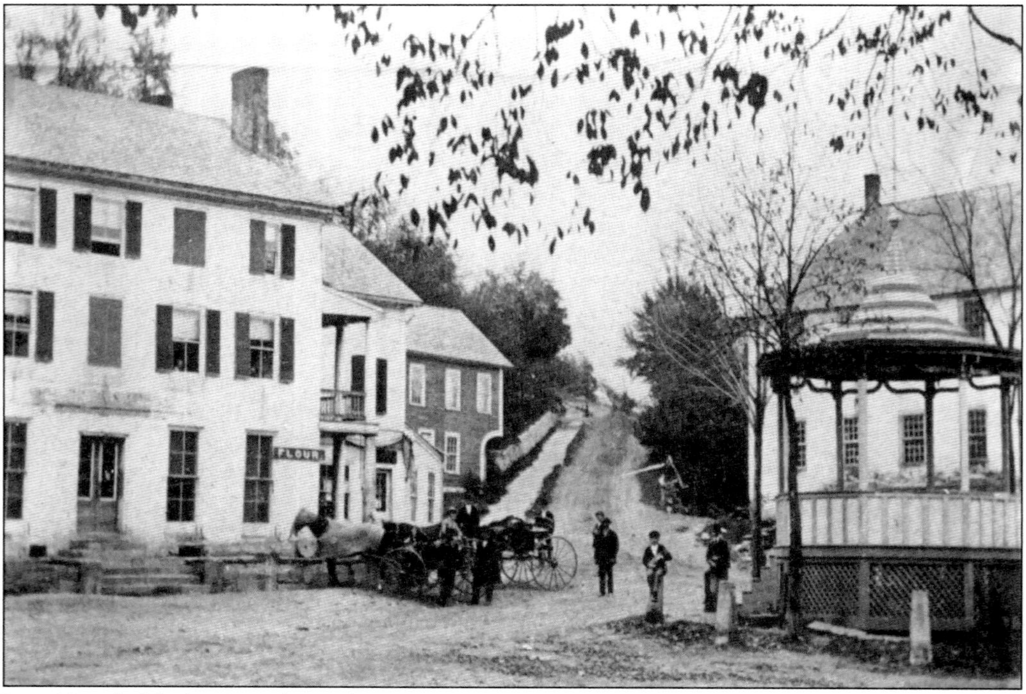


Local lore holds that Jonathan Chase had a general store in 1815 in the building on the left. Isaac Fisher, who lived across the street in what became the Sparrow Block, erected the building for his daughter Sukey, who married Chase in 1818. Chase was known to take a four-horse team to Boston twice a month for supplies while Sukey ran the store. (WGS.)

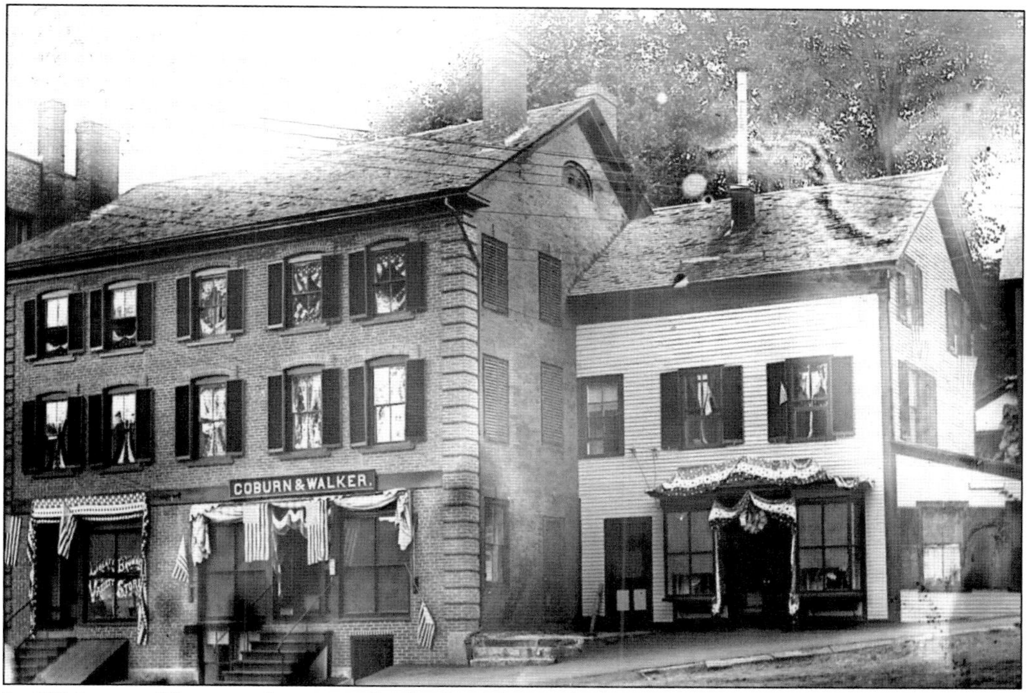


In 1894, some 80 years after it was built, the Chase Block was remodeled. The front was refaced with new bricks, and quoins were added down the sides to dress it up. The side porches were removed, and large shop windows replaced the small windows in front. This remodel, however, would only last until 1903, when it would be redone again. (WGS.)

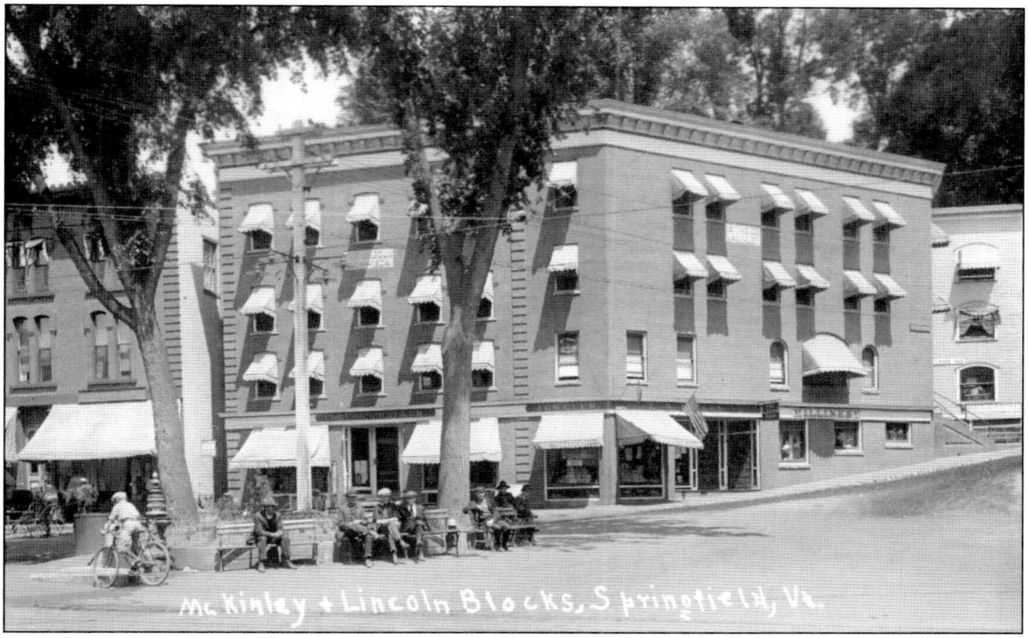


The Chase Block, remodeled again in 1903, ended up as this four-story, flat-roofed block that incorporated some of the buildings to the right. The building would be renamed the Lincoln/McKinley Block in honor of the two assassinated presidents. The safety bicycle (both wheels the same size) became vogue in 1893. Notice the lovely park in front of the building. (STL.)

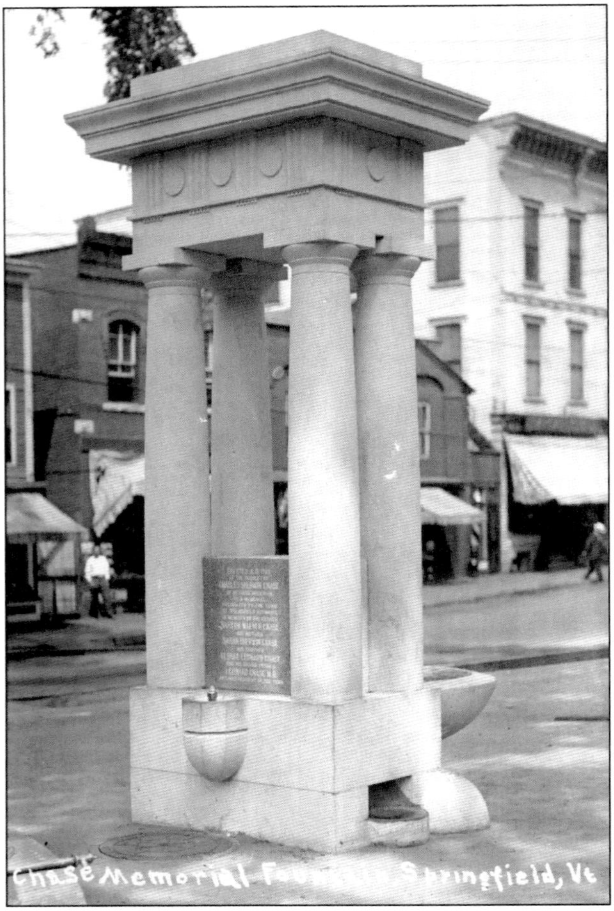


This fountain was erected in 1913 in memory of Dr. Leonard Chase (brother of Jonathan) by his children from Detroit, Michigan. Dr. Chase was not only an eminent physician, he was involved in establishing the seminary, the library, and organizing the two banks. Unlike most other memorial fountains, this one provided drinking water for people, horses, and small animals. (STL.)

The large white building to the right of center is the Fairbanks Block. Though a fairly typical late-19th–century building, its story is not so typical. Asahel Fairbanks, a very successful farmer well known for raising prize oxen, constructed this building in 1889 when he was 82 years old. It is now known as the Odd Fellows Block and is located between the W.T. Grant (SEVCA) Building and the brook. (STL.)

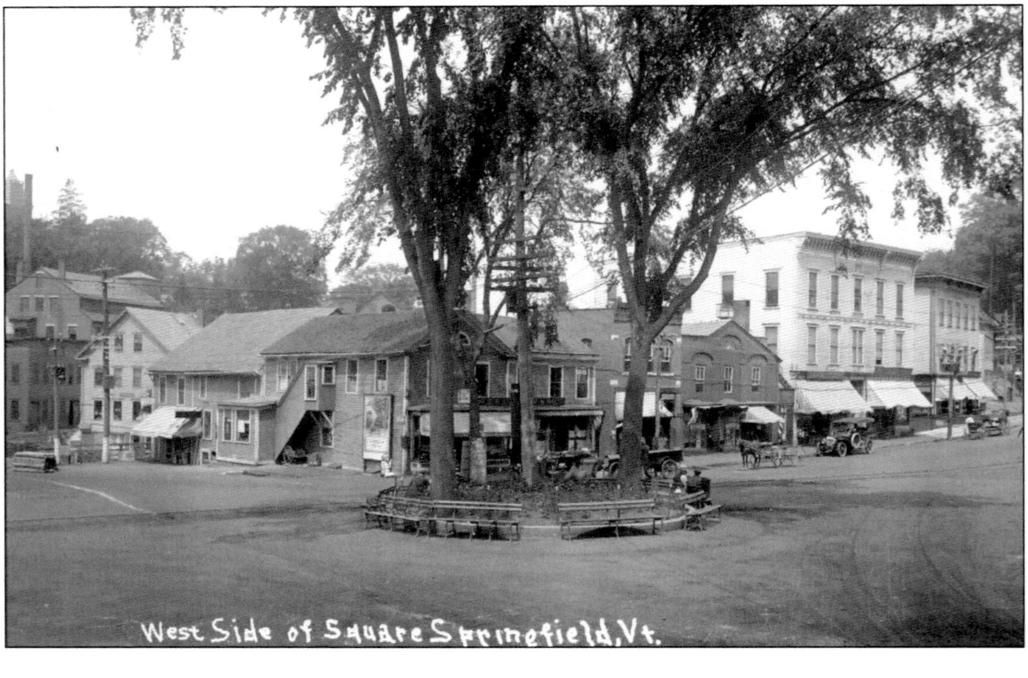

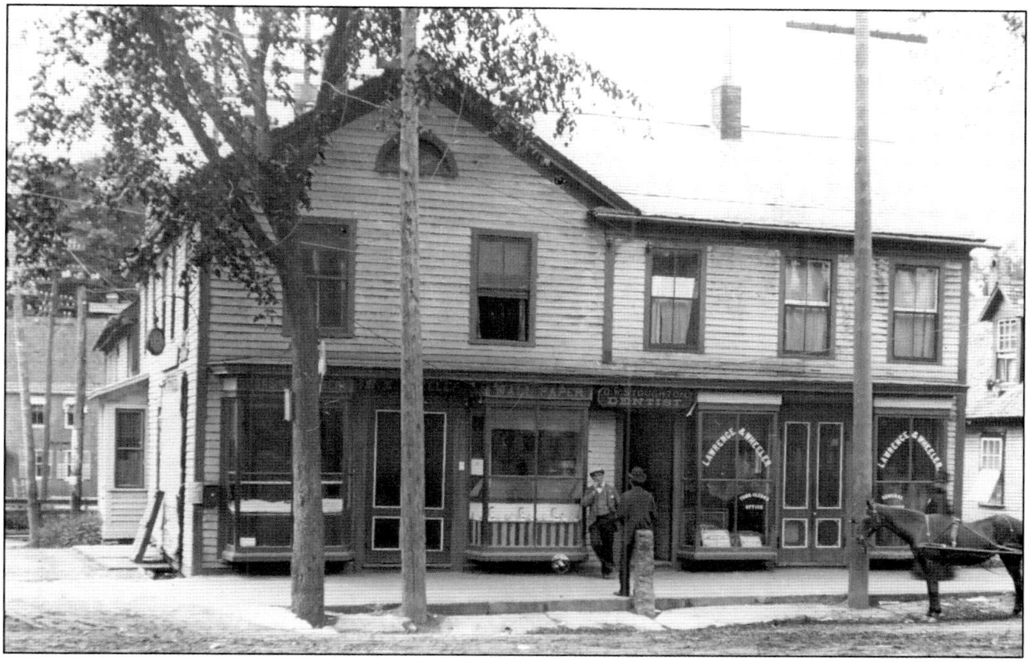


The original building on this site was a spinning-wheel shop. Then the first store was opened in 1806, and in 1849, it became a drugstore. The addition on the right would house a law office and, later, the Lawrence & Wheeler offices. This building would see a great many firsts, including the soda fountain in 1892 and the 50-line telephone switchboard in 1895. (WGS.)

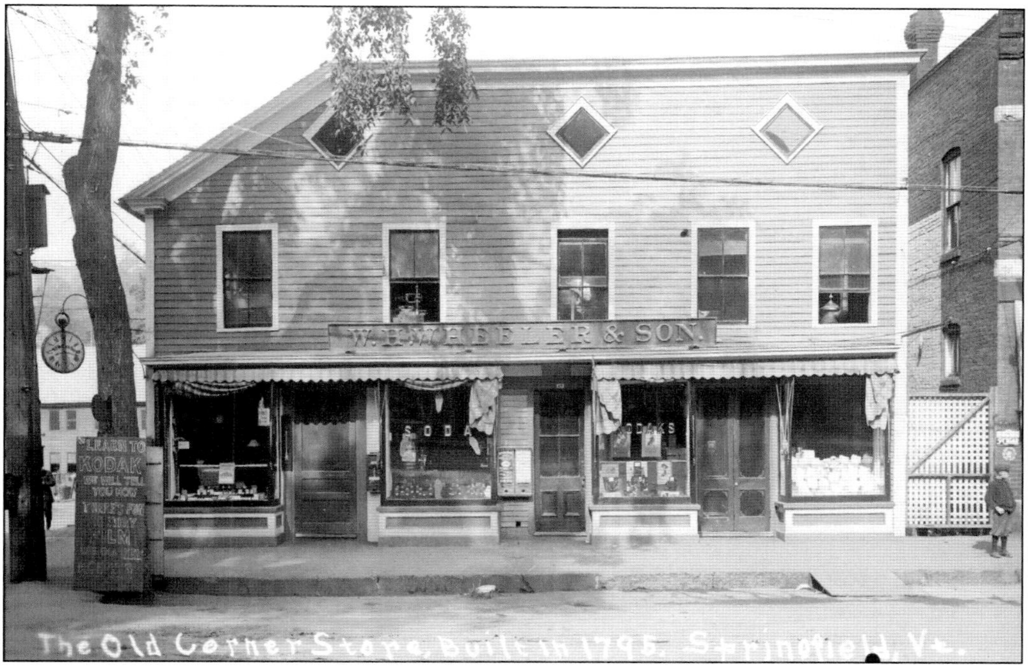


The Wheeler & Son sign was put up in 1901. Around 1912, the building was remodeled, adding a third floor to the right side and the fancy diamond-shaped windows. In October 1922, there was a fire that consumed the music room on the second floor, destroying Grafonolas and Edison, Victrola, and Columbia records. In 1923, a tearoom was added. (STL.)

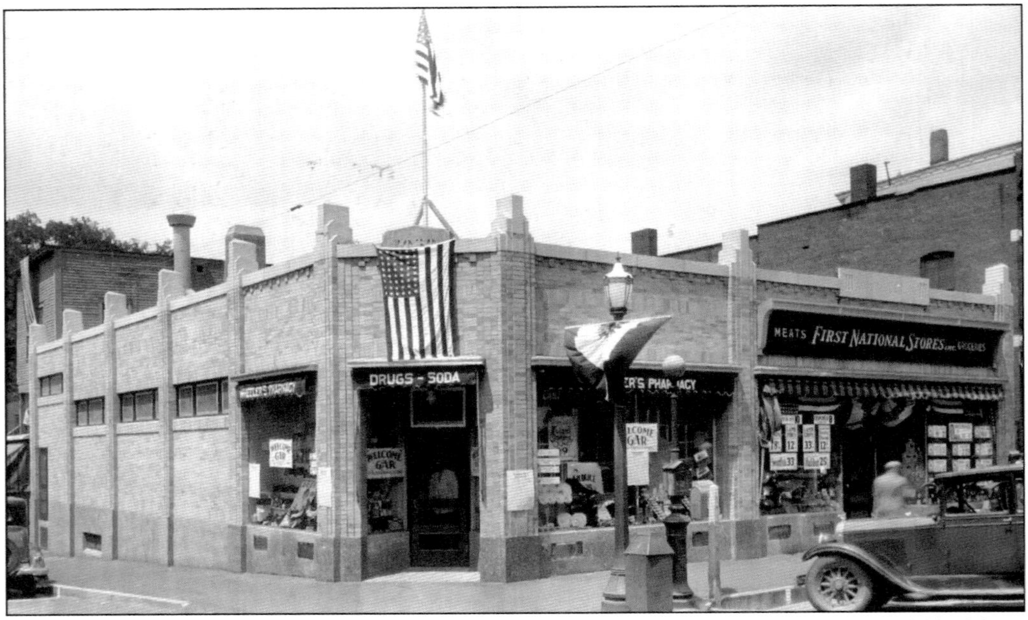


Even with the remodeling, Fred Wheeler finally had enough of his more than 100-year-old building. In 1932, he tore the whole place down and built a new, modern pharmacy with separate retail space next door. The extra space would, in time, be occupied by First National Grocery Store, the Ten Eyck Restaurant, and in 1949, Lloyd's Restaurant, among others. Wheeler's Pharmacy finally closed its doors in 1996. (STL.)

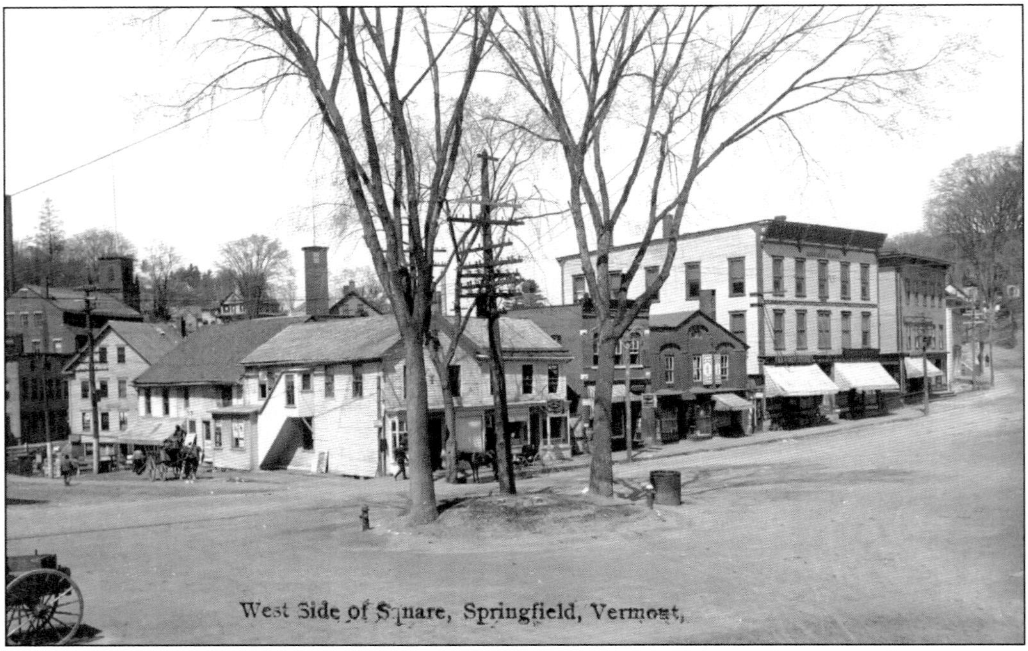


This photograph was taken before 1901. Beginning with the building on the corner, the structures on Main Street are, from left to right, the first Wheeler's Store, the Stiles Building (with the left side addition), the Odd Fellows Block, and the Brown Block. Notice the fire hydrants on either side of the park. James Hartness was so upset when some of the trees on Main Street were cut down that he ran for the post of tree warden. (STL.)

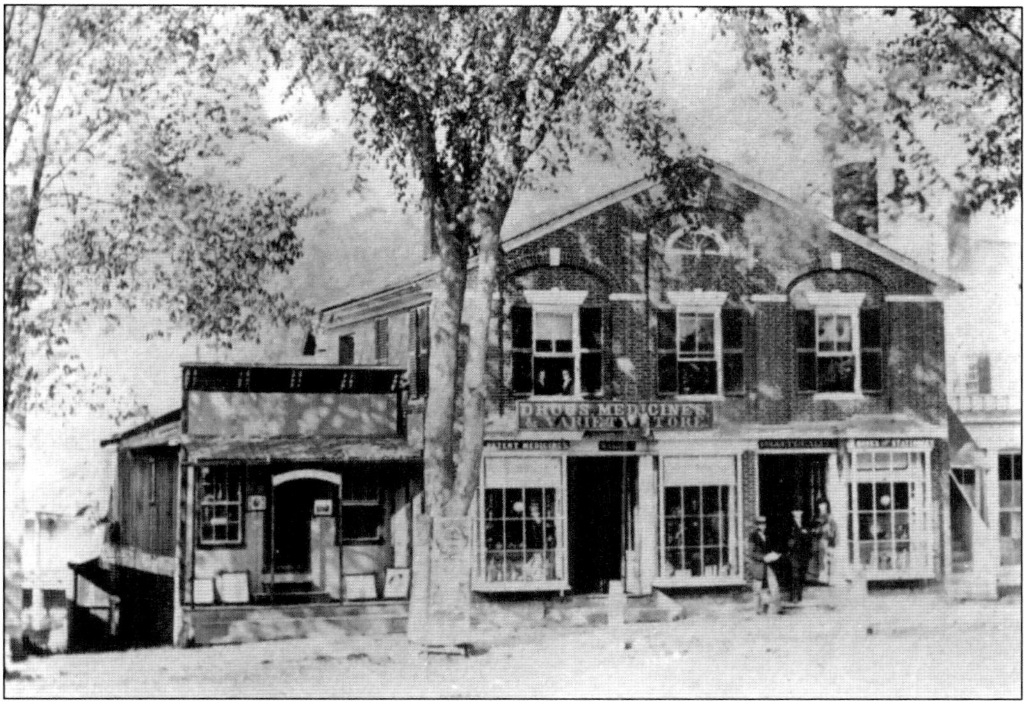


This building is in the earliest photographs of the west side of Main Street, but very little is known about it. It may have been built by the Honorable William M. Pingry, who left Springfield in the late 1830s. At the time of this photograph taken in the mid-1800s, it was a drug and stationery store. The building on the left was later replaced with a two-story addition. (WGS.)

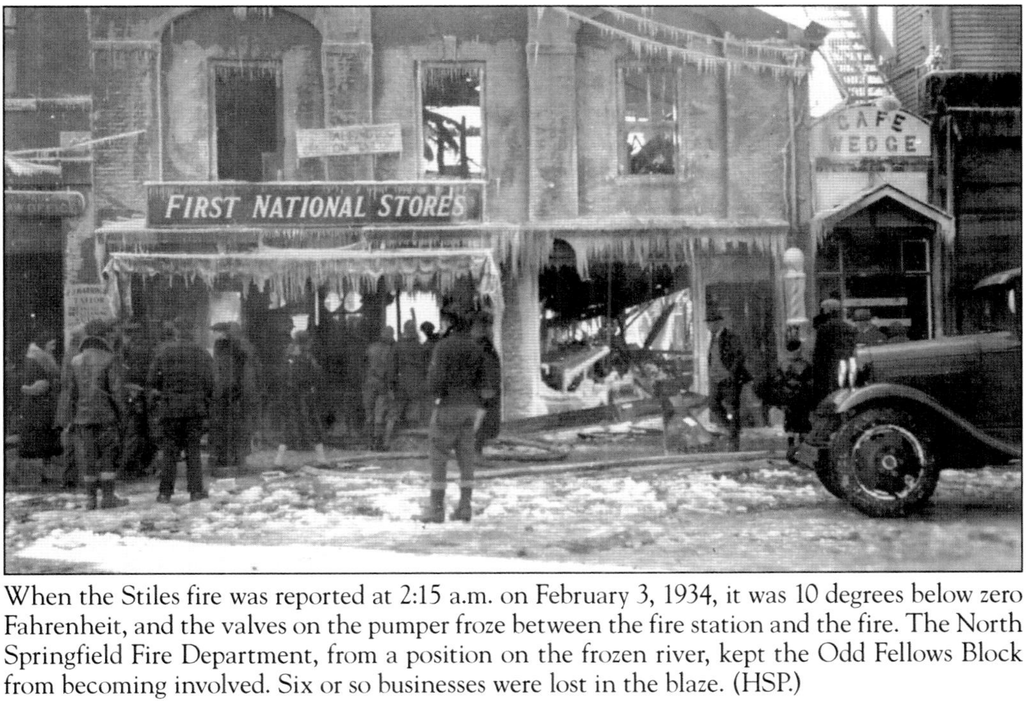


When the Stiles fire was reported at 2:15 a.m. on February 3, 1934, it was 10 degrees below zero Fahrenheit, and the valves on the pumper froze between the fire station and the fire. The North Springfield Fire Department, from a position on the frozen river, kept the Odd Fellows Block from becoming involved. Six or so businesses were lost in the blaze. (HSP.)

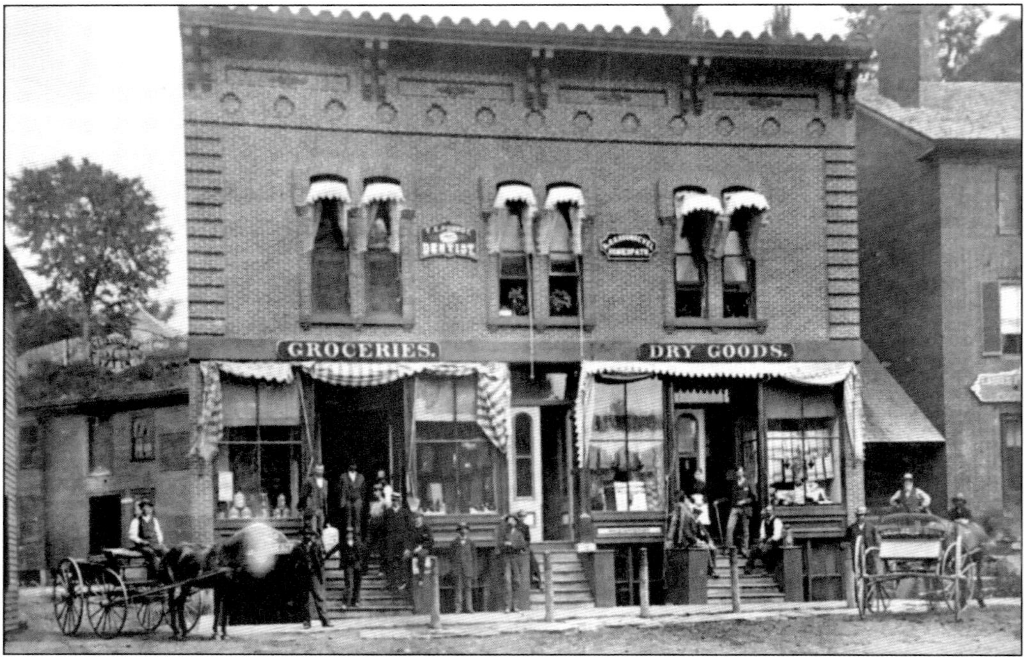


The Leland/Ellis building has had an interesting past. It started off as the two-story Randel Building erected around 1867. Later, Charles Leland bought it and ran a dry goods store on the first floor. He added a third floor as living quarters, which later became rental rooms. The biggest change came in 1931 when the Capitol Theater converted the right side of the store into a movie house. It opened in January and closed in June of that year. It sat vacant until the Avon Theater was opened in 1936. The left side was converted into a restaurant, a flower shop, and, later, a beauty salon. The Avon Theater would become the Ellis Theater and then the Springfield Theater. In 2008, a fire destroyed a large part of the block. Restoration plans are underway. (Above, STL; below, BL.)

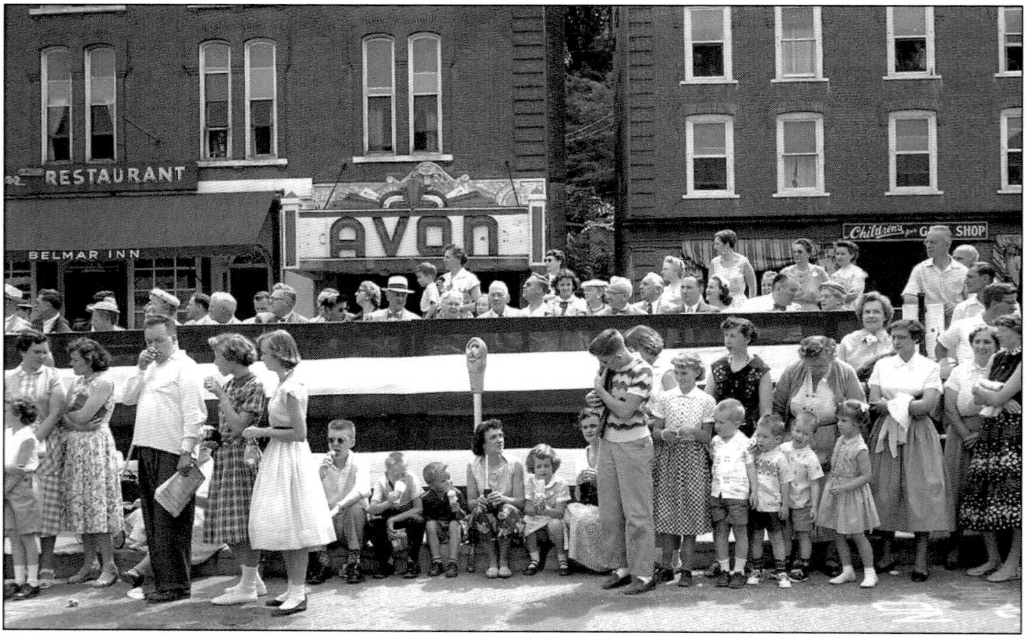

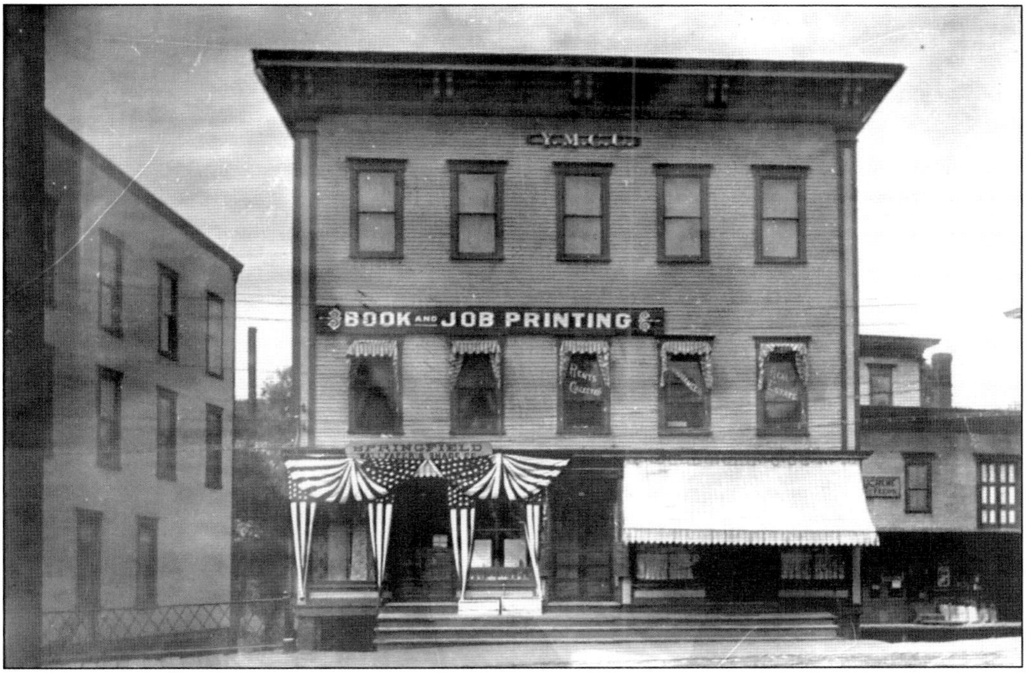


This is a photograph of the Brown Block that was moved from the east to the west side of Main Street to make room for the Adnabrown Hotel. For a while, it was known as the YMCU Building, but more recently it was known as the Bakery Building. It is difficult to recognize today because the five windows on the third floor are now two and a false first-floor front was added. (STL.)

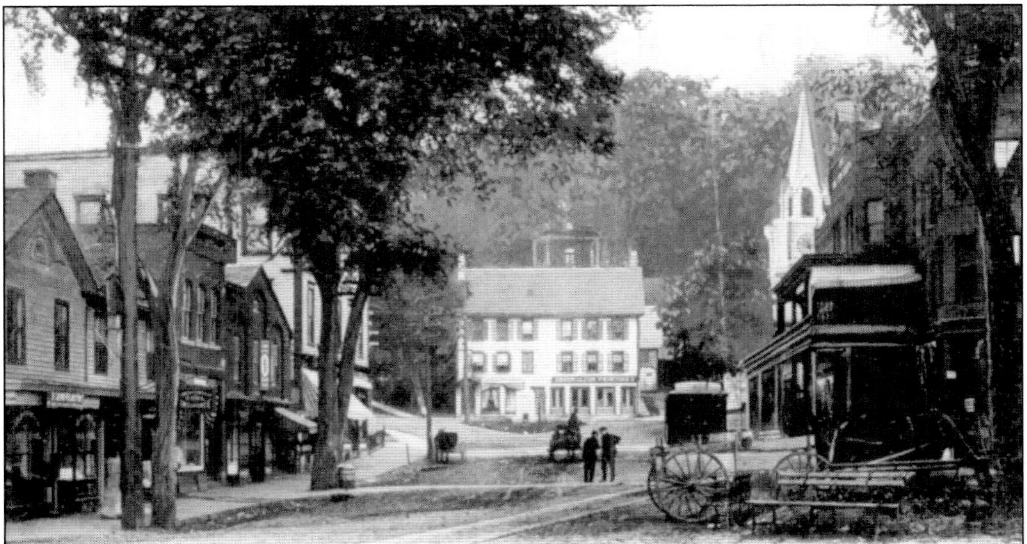


The building at the head of the square has a stone in the attic inscribed "1833." Built as a duplex, in *Folklore of Springfield, Vermont*, Eva Baker described it as "the first block of any size." She continued, noting that "men came from far and near to see it and pass their opinion on its merits." It is known as the Dillon or Tarro Block. This image was taken before 1901. (HSP.)

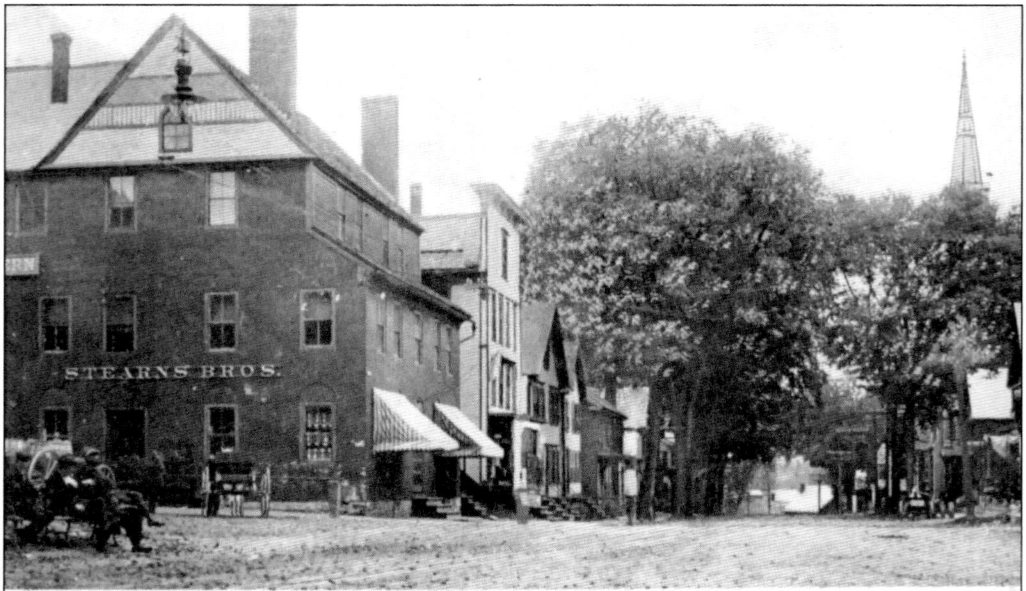


*Springfield Vermont, Main Street Showing Commonwealth Block*

On the southeast side of the square, the brick building on the left was originally erected as Isaac Fisher's mansion in the early 1800s. By mid-century, it had already undergone many changes, with many more to come. Next to it is the Lawrence & Wheeler Building. The remaining buildings on the left would be torn down or moved back to make way for the Bank Block in 1907. (STL.)

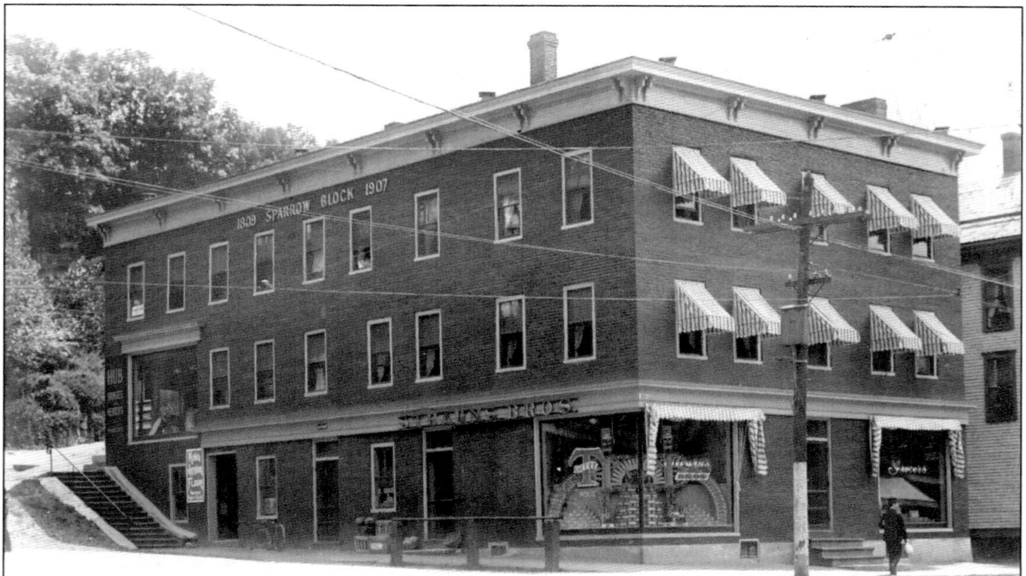


Before 1907, this building was known as the Tontine or Commonwealth Block. In 1907, however, William Sparrow did some major remodeling—the front was straightened, the roof flattened, and the front entry altered—and it was renamed the Sparrow Block. In 1909, it would house Dreamland, the first movie theater in Springfield. (STL.)

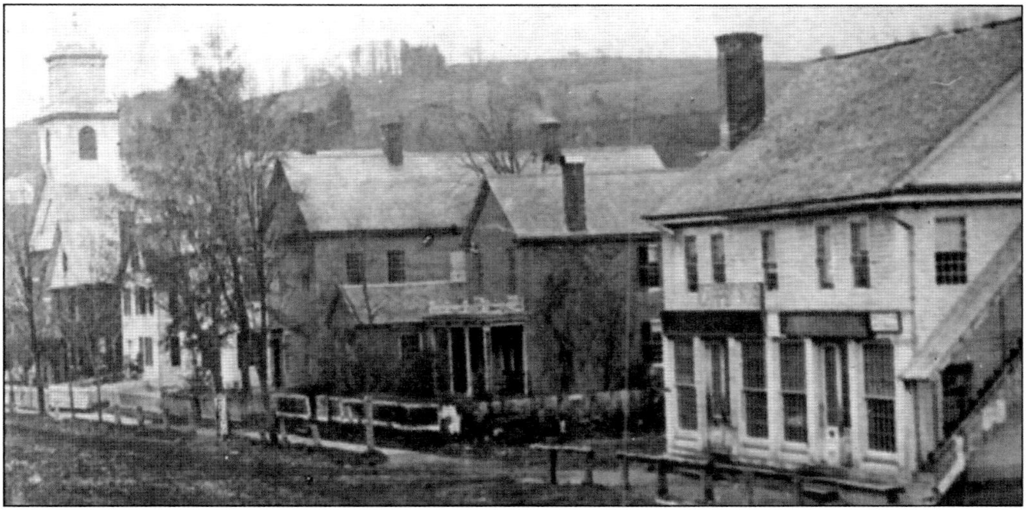

Erected in the early 1800s, the building on the right (above) is the Brooks Store. It stood on the corner of Main and Park Streets where the Woolson Block now stands. The other buildings are (right to left) Judge Closson's home and small law office, two buildings that were later torn down, the W.H.H. Slack home (Masonic Temple), and the Congregational Church. In 1867, to make room for the Woolson Block, Minor Newton purchased the Brooks Store and moved it to Valley Street (below). He placed it over the brook just below Pleasant Street. It was used as a store and was later sold to the Springfield Grange. It was torn down in 1987. A selectman recalled that the porch on Valley Street took a beating from the snowplowing because the building was so close to the road. (Above, WGS; below, BL.)

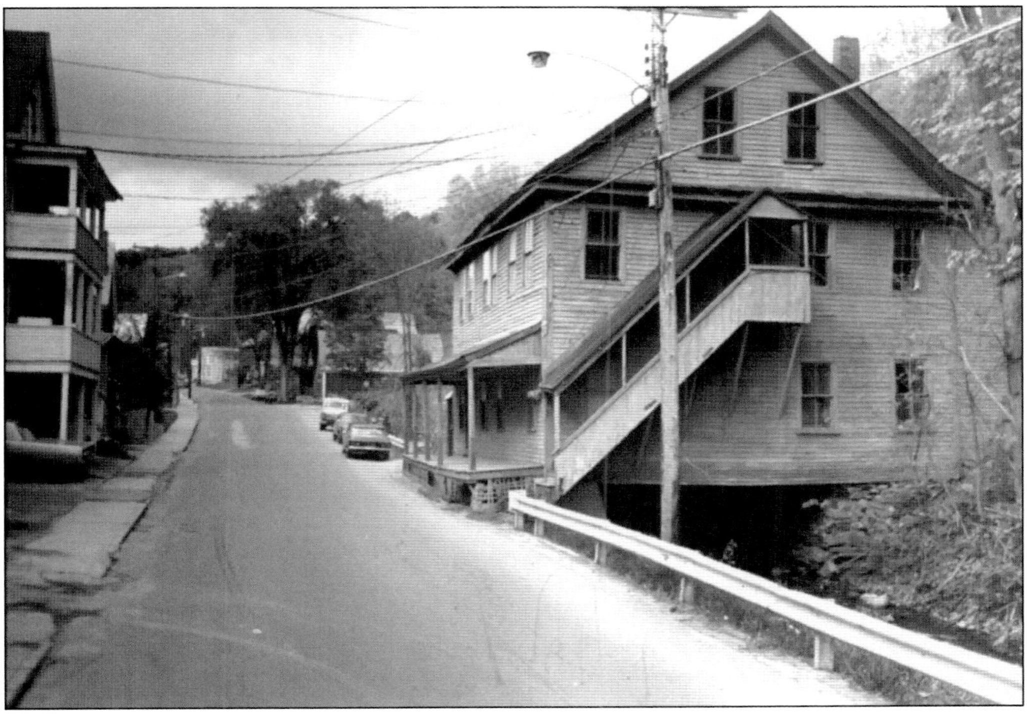

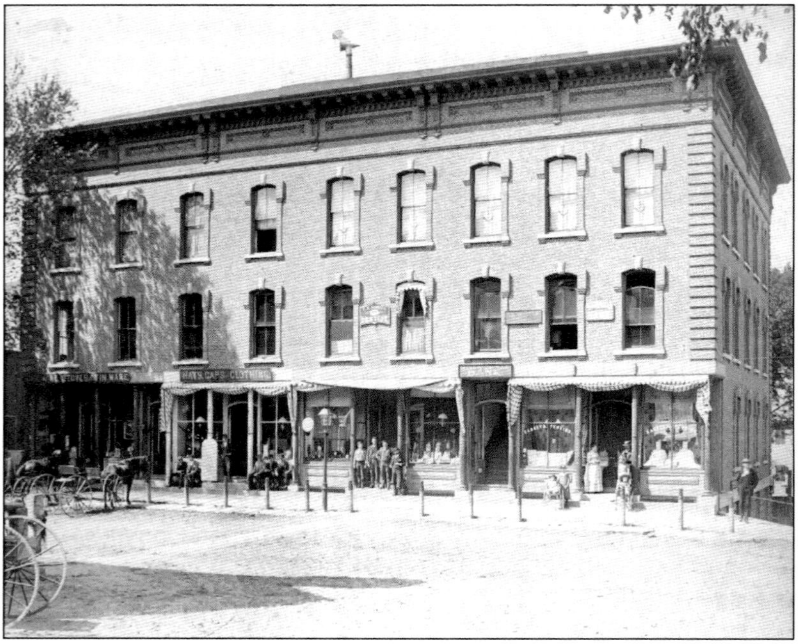

Built in 1869, the Woolson Block contained stores on the first floor and offices upstairs. A tin shop, dry goods store, and drugstore are on the first floor in this photograph taken before 1901. The First National Bank and Dr. F.E. Pomroy, a dentist, are on the second floor. The Masons are on the top floor. (STL.)

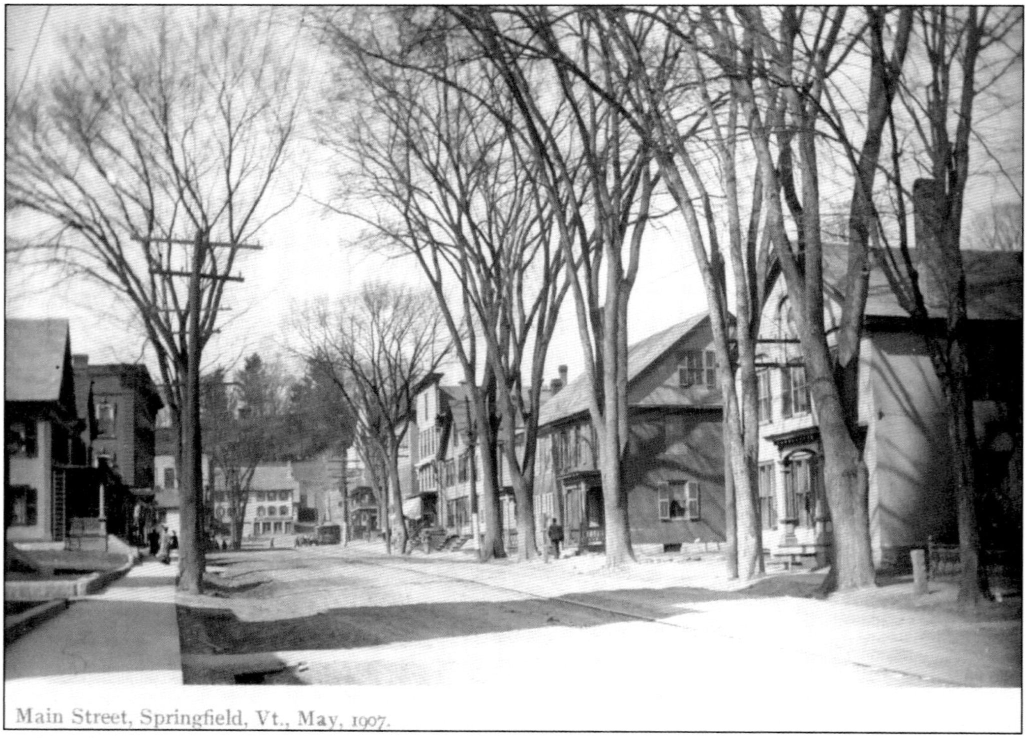


Before the early 1900s, Main Street was lined with trees. It had small businesses and homes with picket fences. The construction of the Bank Block would change Springfield's image from that of a small village to that of a growing town. The elm trees were allowed to stand for a while, but soon they would be considered a nuisance and cut down. (STL.)

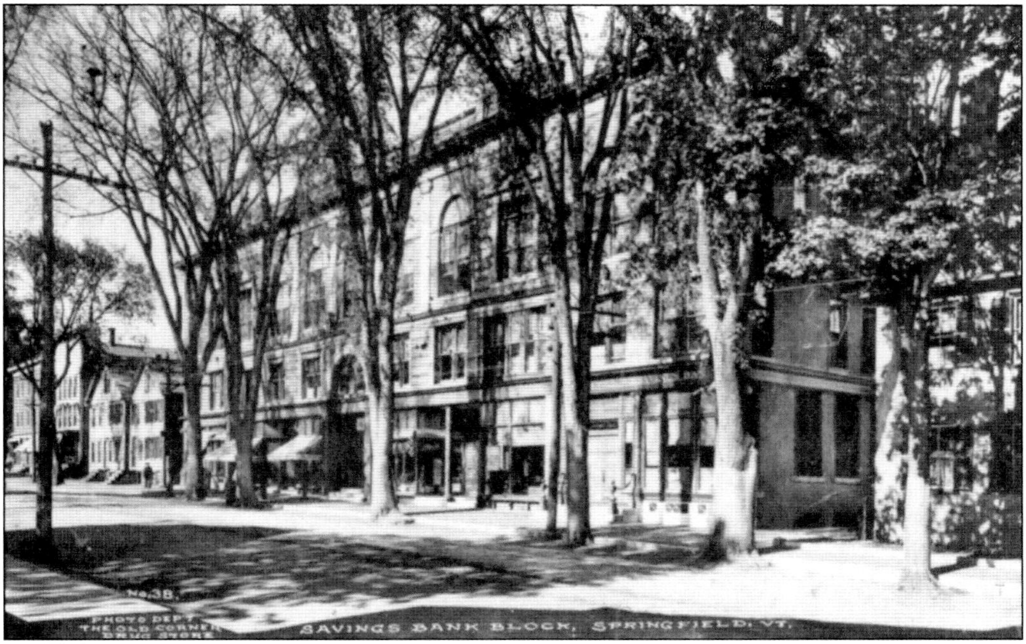

Originally one of the prettiest on Main Street, this building was a product of the combined efforts of the First National and the Savings Banks. Completed in only seven months, it opened July 1, 1908. Over the years, it has housed the town offices, the police station, the telephone company, the post office, several fraternal organizations, professional offices, and a variety of stores and restaurants. (STL.)

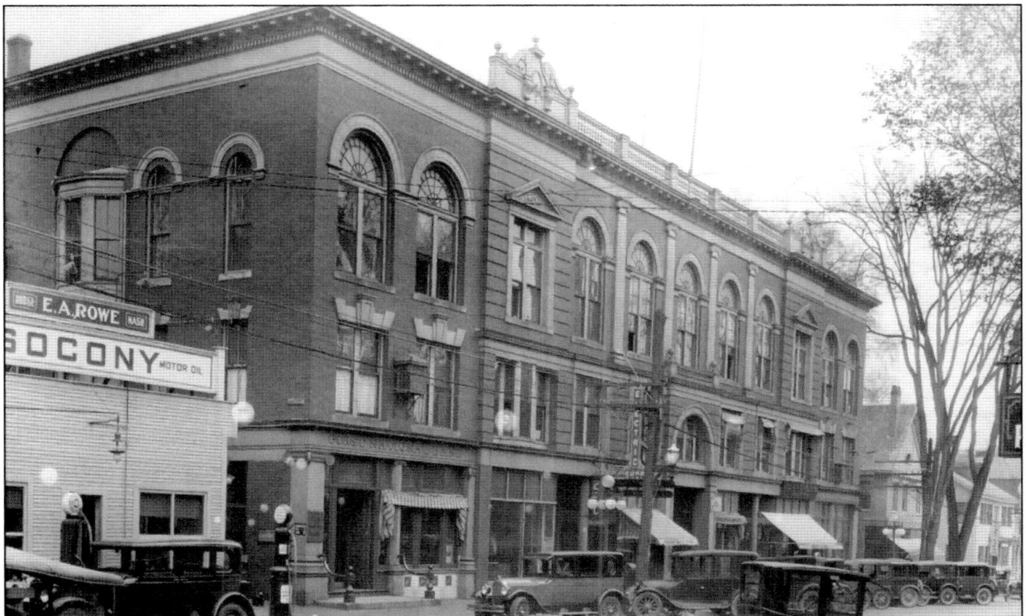

This photograph was taken after 1924. Already, many of the trees had been cut in front of Bank Block. Along with the town and telephone offices, there was a shoe store and a variety store. The building on the left became a gas station. Gasoline was 23¢ per gallon, plus a 2¢ tax. It was replaced with a new building containing smaller stores and finally became the bank parking lot. (STL.)

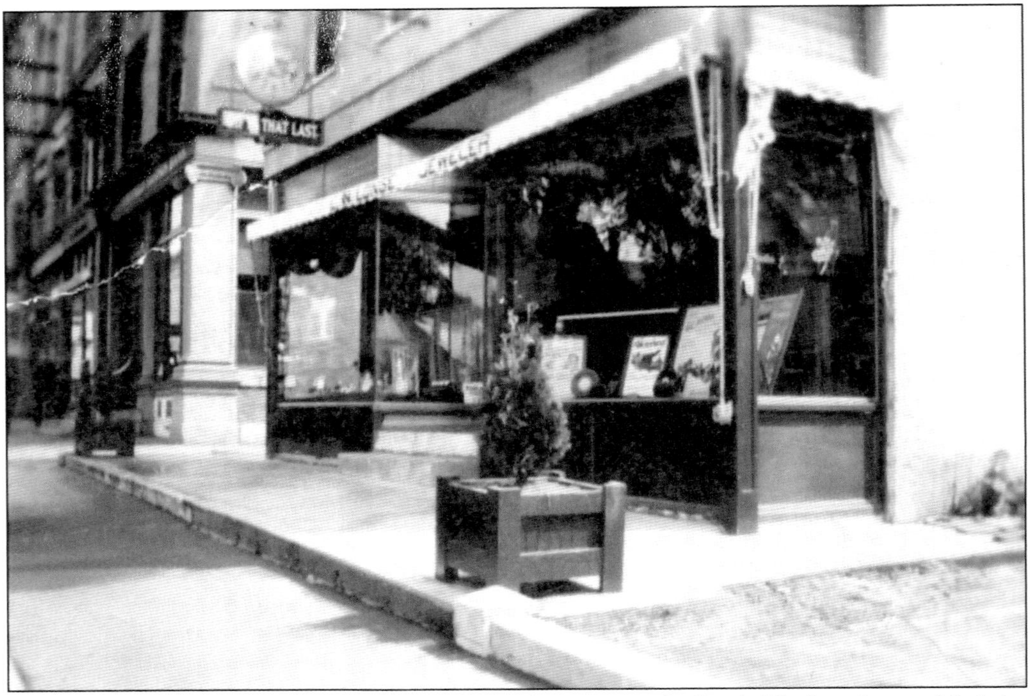

In 1924, L.N. Longe bought the dwelling next to his store in the Bank Block and remodeled the front portion into a modern jewelry and music store (pictured). He ran this store for five years and then sold the property to a Boston company that planned to build a modern business block. The new block would become the A&P/J.J. Newberry building. (STL.)

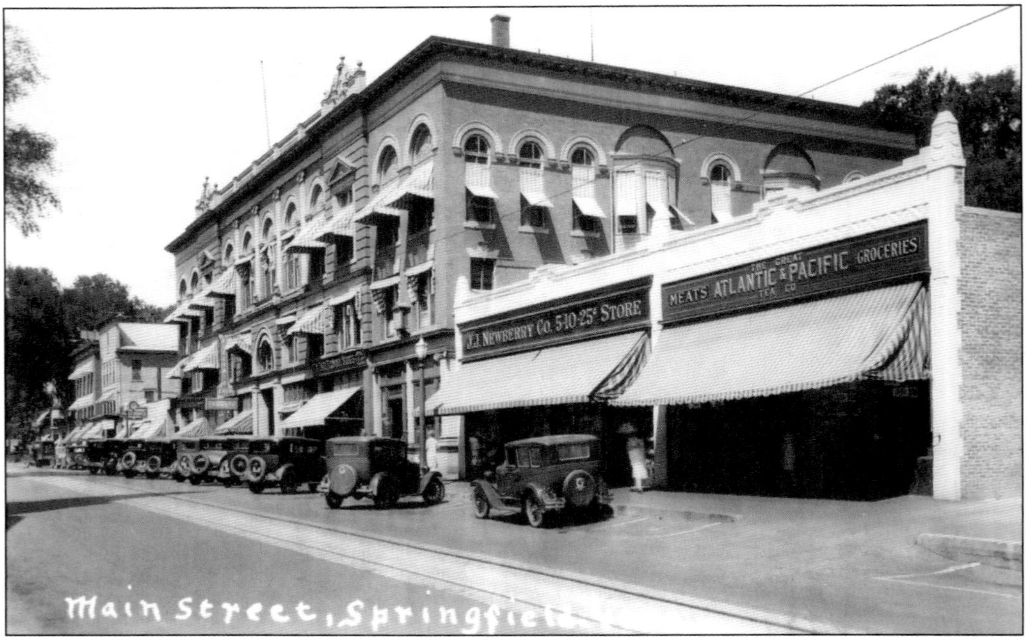


The A&P moved into the right side of the block in September 1929. The store had a meat department and probably four aisles. In earlier days, meat was sold by a butcher out of a meat wagon. J.J. Newberry's, a novelty store chain, moved into the left side. For 40 years, these two stores would serve as the backdrop for many brides arriving at the Congregational Church. (SAB.)

# *Three*

# Around Town

This chapter introduces many places of interest, including schools, public buildings, churches, prominent homes, intriguing old street scenes, and overviews of the town. As it would take 15 photographs to show all of the schools and the changes made to them before 1970, only the five oldest are covered in this chapter.

Public buildings included are the library, hospital, town hall, and fire station. The prominent homes include Closson, Ellison, Hartness, Smiley Manse, Williams, and Woolson, as well as two apartment buildings. The Miller Art Center, now home to the Springfield Art and Historical Society, was built between 1861 and 1865. The beautiful, two-story brick mansion was extensively remodeled in 1917 when the pillars and porch were added. In 1956, it was donated to the town by Grace and Edward Miller to be used as an arts center.

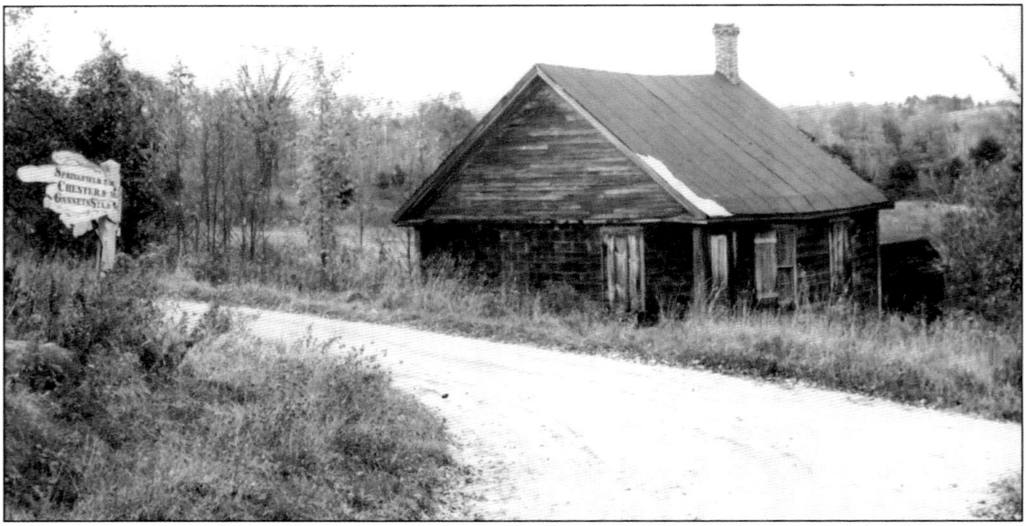

The first schoolhouse was built in 1772. Eureka, the second schoolhouse, was originally located at the corners of Eureka, Barlow, and Greeley Roads. According to Hubbard and Dartt's *History of the Town of Springfield, Vermont*, it would usually hold 80 students, but sometimes it held as many as 125. What remains is part of the Information Center located on Route 10 coming into town. (MAC.)

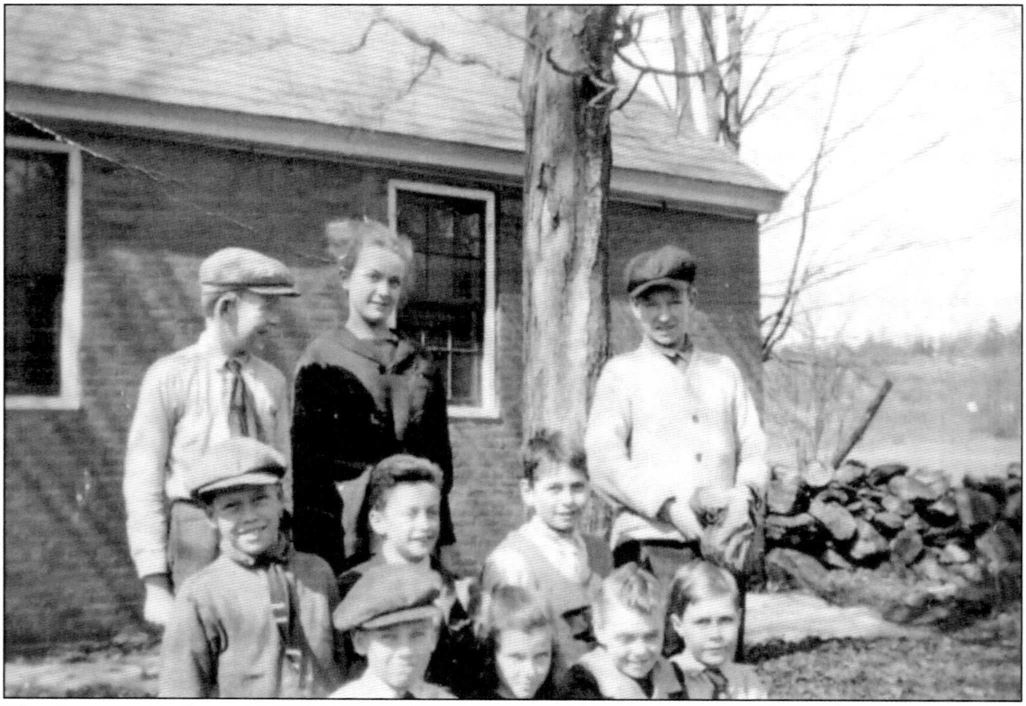

The Spencer Hollow School is located on the corner of Skitchewaug Trail and Spencer Hollow Road. It is claimed to be the oldest brick schoolhouse in the state. In *Folklore of Springfield, Vermont*, Eva Baker notes that the school was built around 1781. This photograph was taken in the early 1920s. Pictured here, left to right, are (first row) Richard Whitcomb, Margret Whitney, Janet Hunsdon, and Leslie Reagan; (second row) Marshall Willard, Arthur Wells, and Fred Crary; (third row) Austin Goings, Florence Whitcomb, and Frank Rumrill. (Don Whitney.)

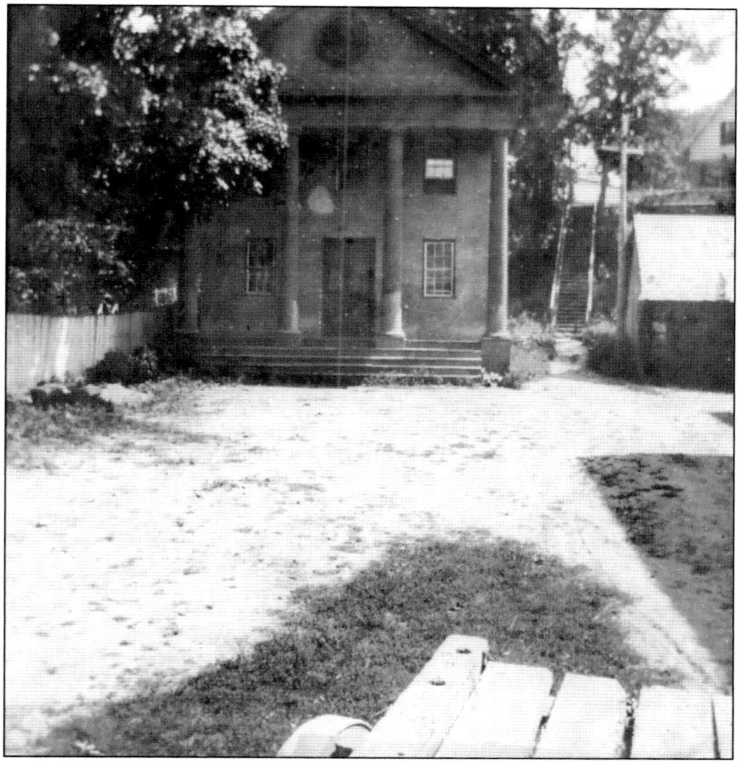

Located on Main Street, the primary grammar school was most likely built before 1833. It was to the left of the town office parking lot (where the Universalist Church was located) and just below the Brookline Apartments. Herbert Ellis, in his childhood recollections, described it as having two grades on each floor and a winding stairway on each side of the vestibule. (WGS.)

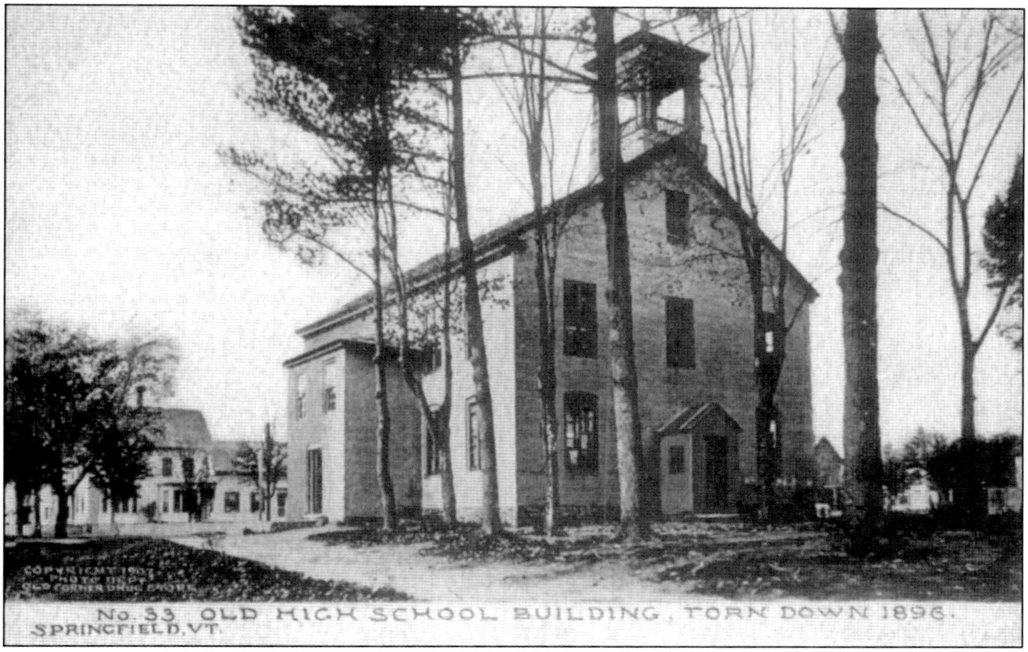


From 1847 to 1866, the older students in town had the benefit of the Springfield Wesleyan Seminary, which was owned by the Vermont Methodist Conference. In 1866, the conference decided to consolidate and close this school. In 1867, after a short interim, the town bought the building for $5,000 and continued to use it for secondary education until 1895, when Park Street School was built. (STL.)

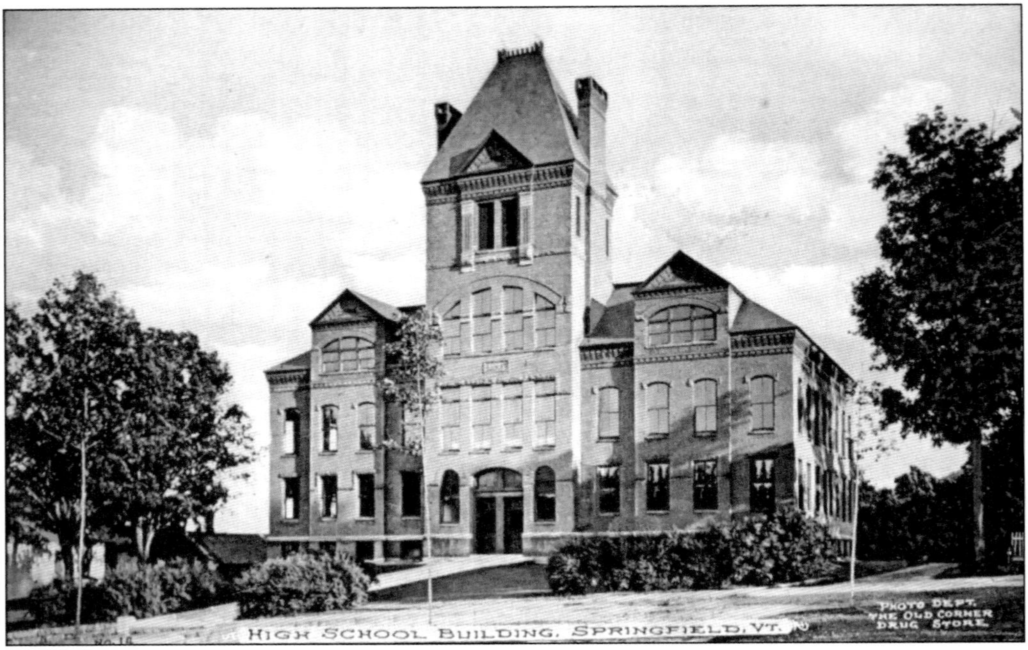


The Park Street School cost $53,500 to build in 1895. The first floor had six classrooms and four teachers' rooms. The second floor held the high school and recitation rooms, a large music hall, and two grammar school and three teachers' rooms. The third floor held the gym, a laboratory, a recitation room, and storerooms. There were closets and storage in the basement. The school had steam heat provided by the shoddy mill. (STL.)

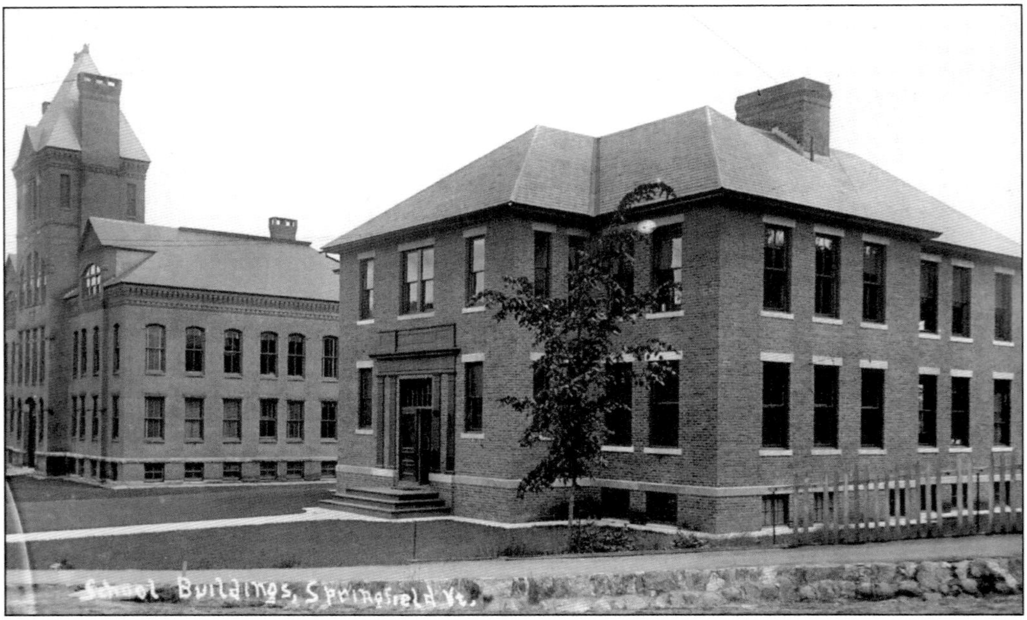


This photograph, though similar to how the building looks today, is misleading. The structure on the right is the Central School, a separate building constructed in 1911 to house the lower grades. Water fountains and flush closets in the basement were hailed as modern improvements. In 1929, it was moved back to make room for the current addition, which also is when the tower was removed from the other building. In 1944, the Central School burned down. (STL.)

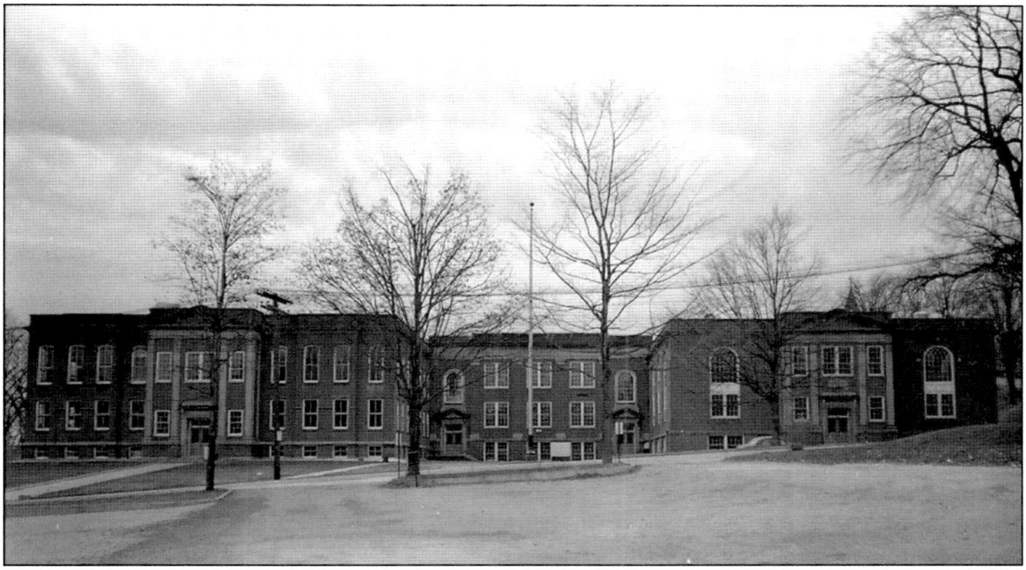

In 1929, the tower was removed, and a large addition was made to the 1895 school building. It included an 800-seat auditorium and a 600-seat gym. It also contained a cafeteria, 20 classrooms (including rooms for physics, chemistry, home economics, and business and cooperative courses), and office space. This addition was often referred to as the new building. (BL.)

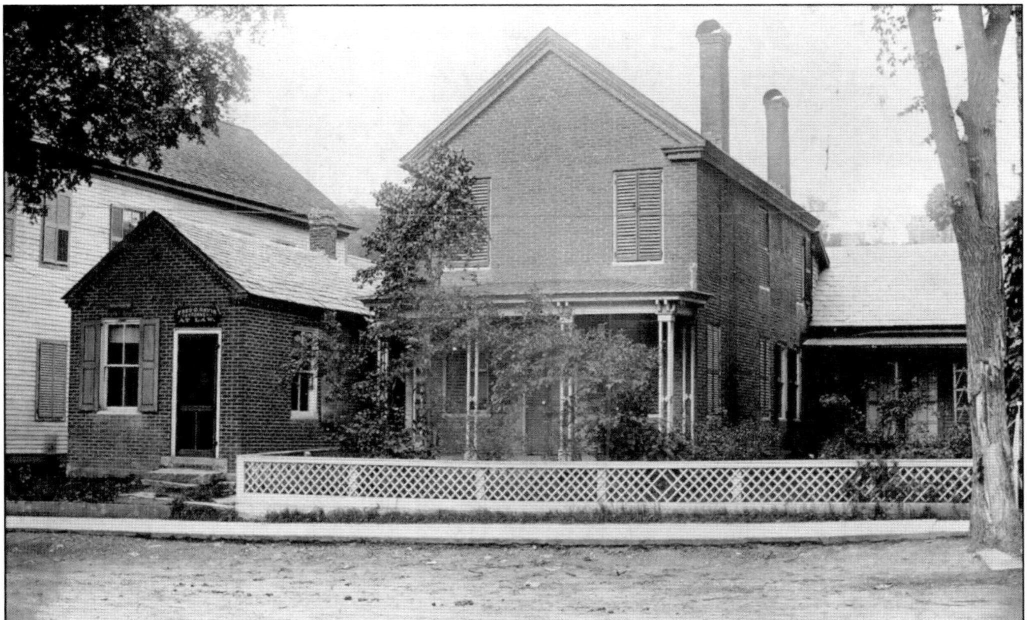

Judge Henry Closson lived in this house next to the Woolson Block and had his law office in the small brick building to the left. He was a prominent man in town and the second of 10 librarians who housed the books. He passed away in 1880. The Sparrow family owned the house when it was torn down to make room for the new 1895 library. (WGS.)

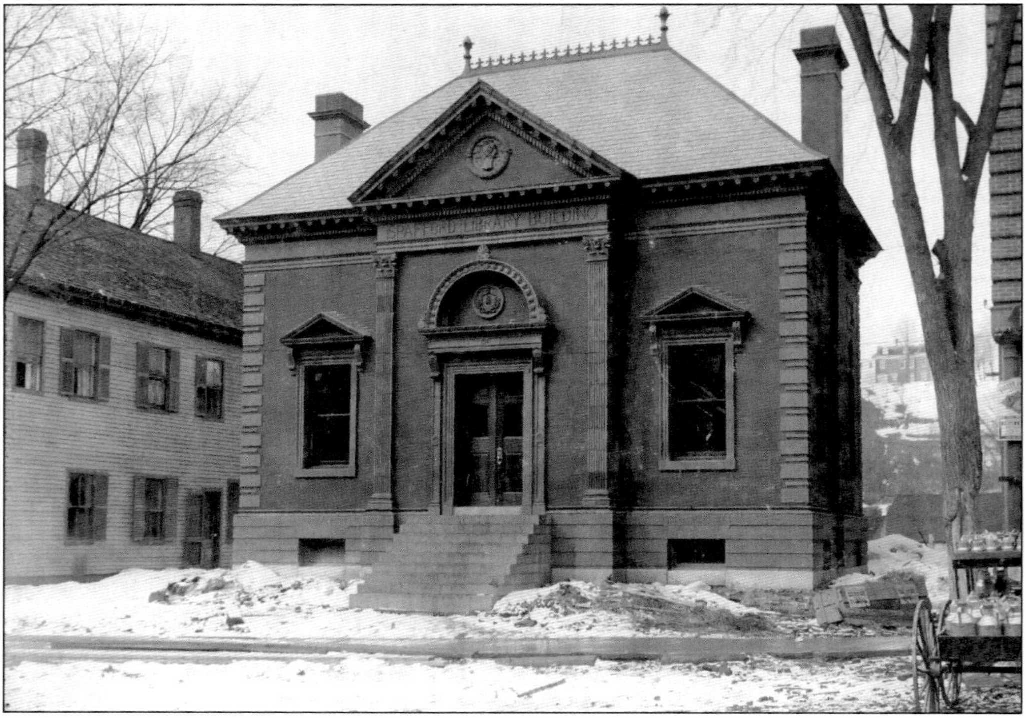


This is not a Carnegie library. When Henry Harrison Spafford passed away in 1893, he left $20,000 to erect and maintain a public library. Built in 1895, it was 40 feet by 56 feet. It contained two reading rooms and held 15,000 volumes. In an ell in the back, it had a beautiful stained glass window made by an unidentified artist. (STL.)

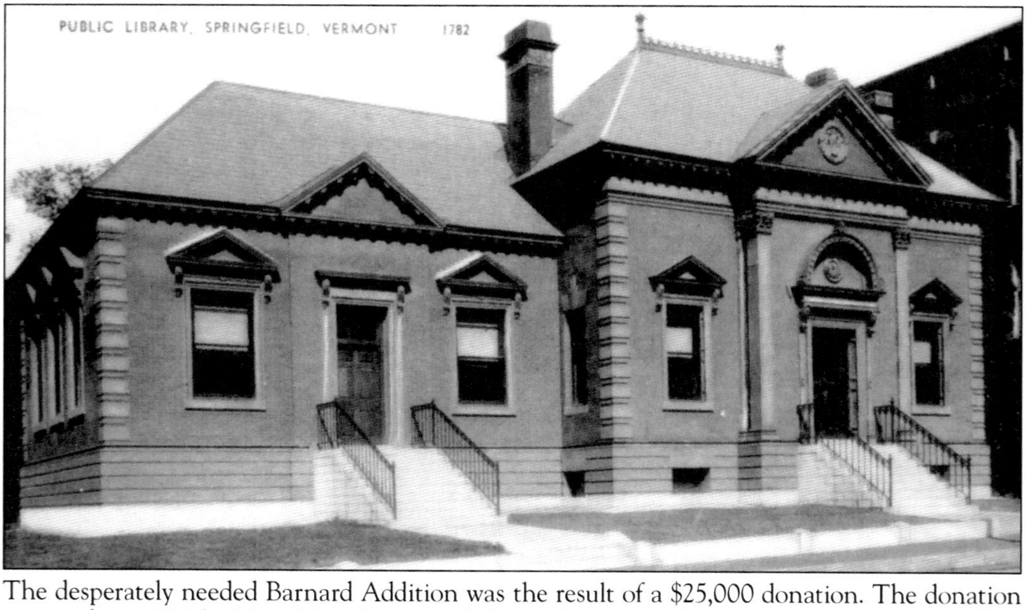


The desperately needed Barnard Addition was the result of a $25,000 donation. The donation was made in 1926 by Mary Eliza Barnard of Grand Rapids, Michigan, in memory of her father, Henry Barnard, who died in 1873. Henry Barnard had been a prominent citizen in Springfield and was instrumental in forming the First National Bank in 1863. Miss Barnard continued to make donations to the library in subsequent years. (HSP.)

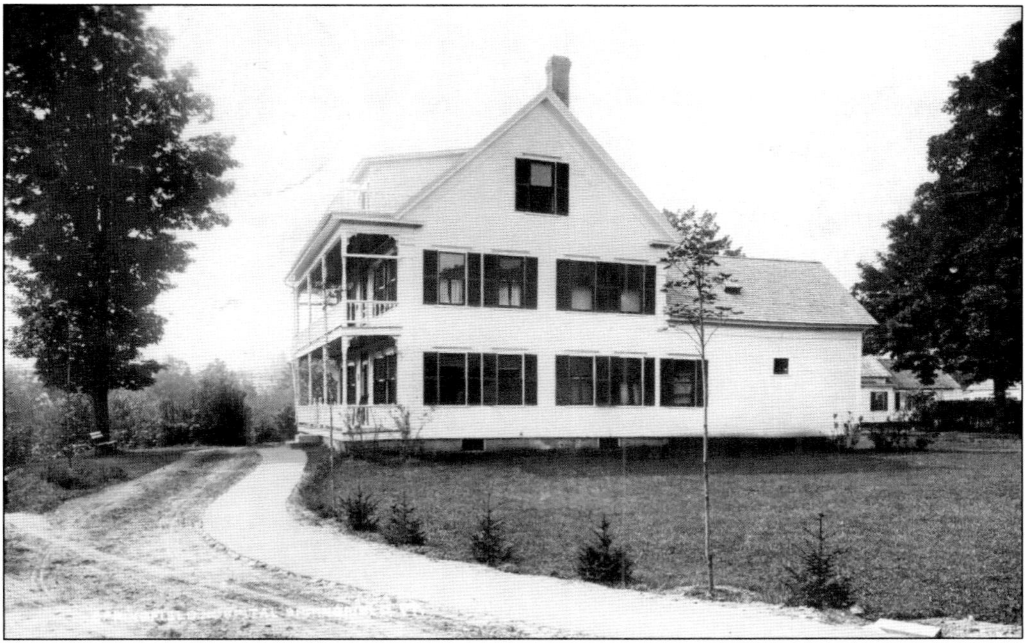

In 1913, the Brotherhood Bible Class of the Methodist Church decided to address the town's lack of a hospital. Because time was of the essence, the Hospital Aid Association decided to lease this 10-room house on Mt. Vernon Street for $25 per month. The hospital opened on February 26, 1914, charging a rate of $12 a week, exclusive of doctors' fees and medicine. (HSP.)

This is the Nahum Charles Dodge Room. It appears that rooms may have been sponsored by various groups. At the time, the *Springfield Reporter* was filled with long lists of items donated by public-minded citizens. These included dishes, vases, tablecloths, towels and washcloths, bedding, and books. Particularly interesting was the entry listing four metal hot-water bottles. (STL.)

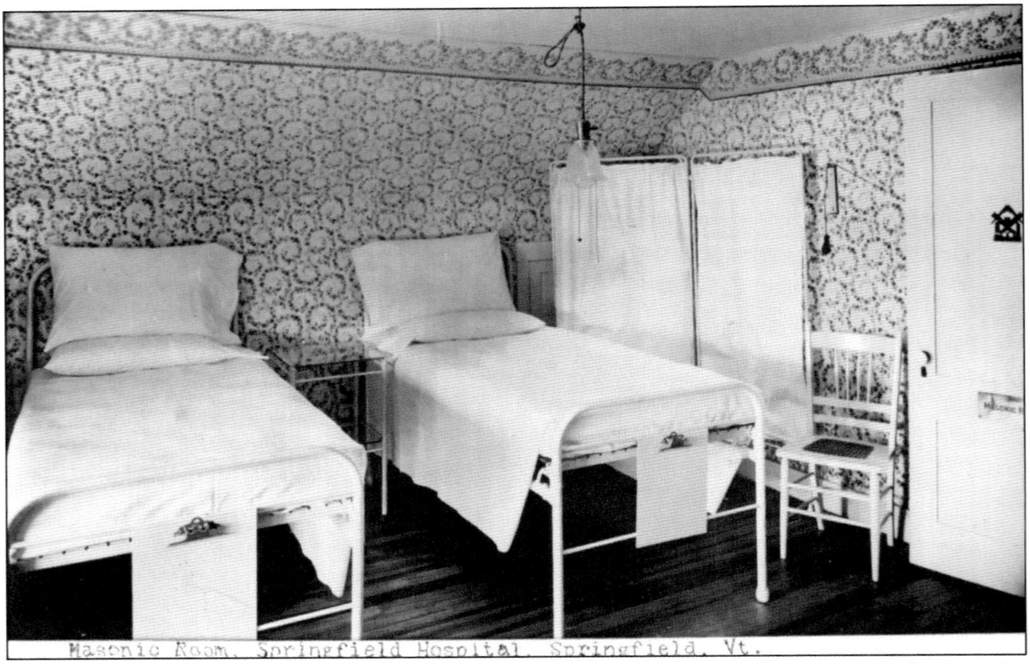


The Masonic Room, with its slanted ceiling, clarified that these three patient room photographs were taken in the Mt. Vernon Street Hospital. Although these photographs show only two beds in each of the rooms, there must have been more in some rooms, given the number of patients cared for at the hospital. (STL.)

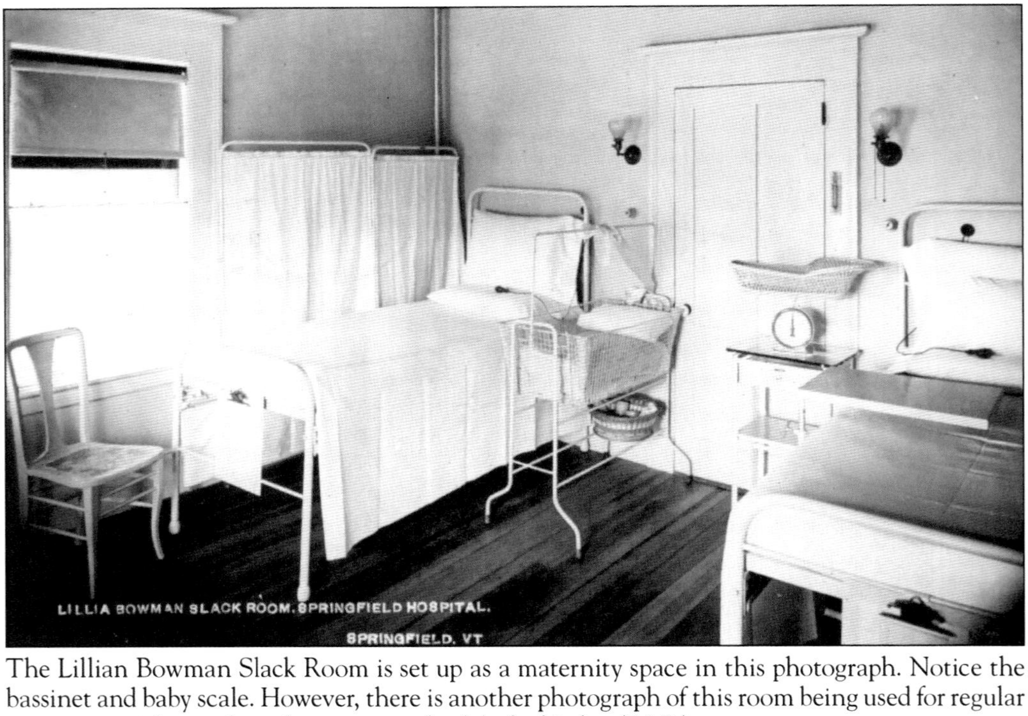


The Lillian Bowman Slack Room is set up as a maternity space in this photograph. Notice the bassinet and baby scale. However, there is another photograph of this room being used for regular care, so it might not have been reserved solely for births. (SAB.)

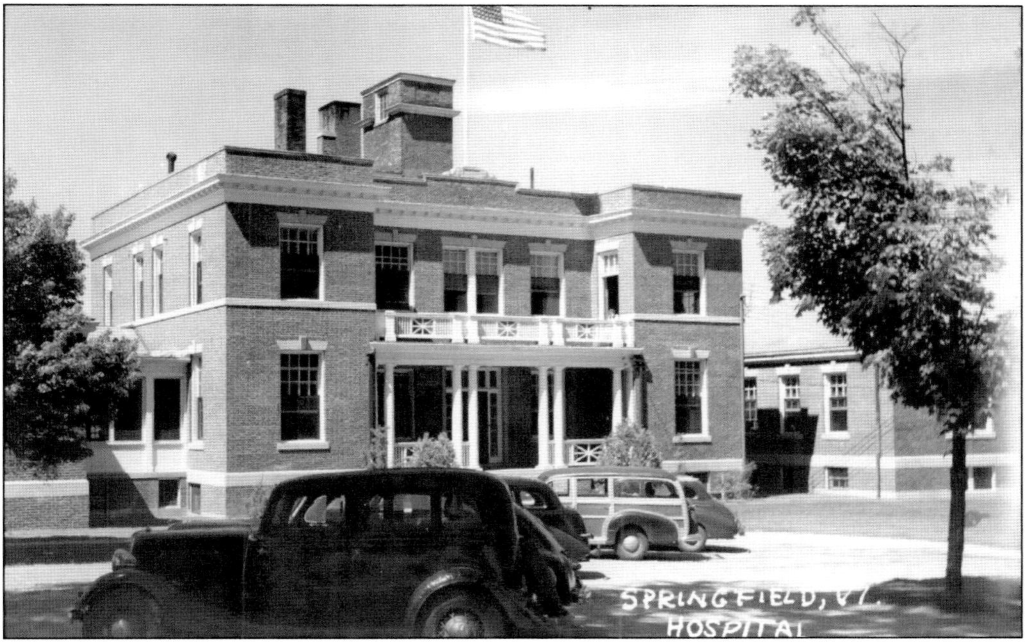


As early as 1917, A.W. LaFountain was on a one-man campaign to build a new hospital. He raised $30,000, and that same year, Henry J. Whitcomb, a former resident, gave the town his summer home and six acres of land off Park Street as a hospital site. It was 1920 before a contract was signed, and the new hospital would not officially open until June 30, 1922. (STL.)

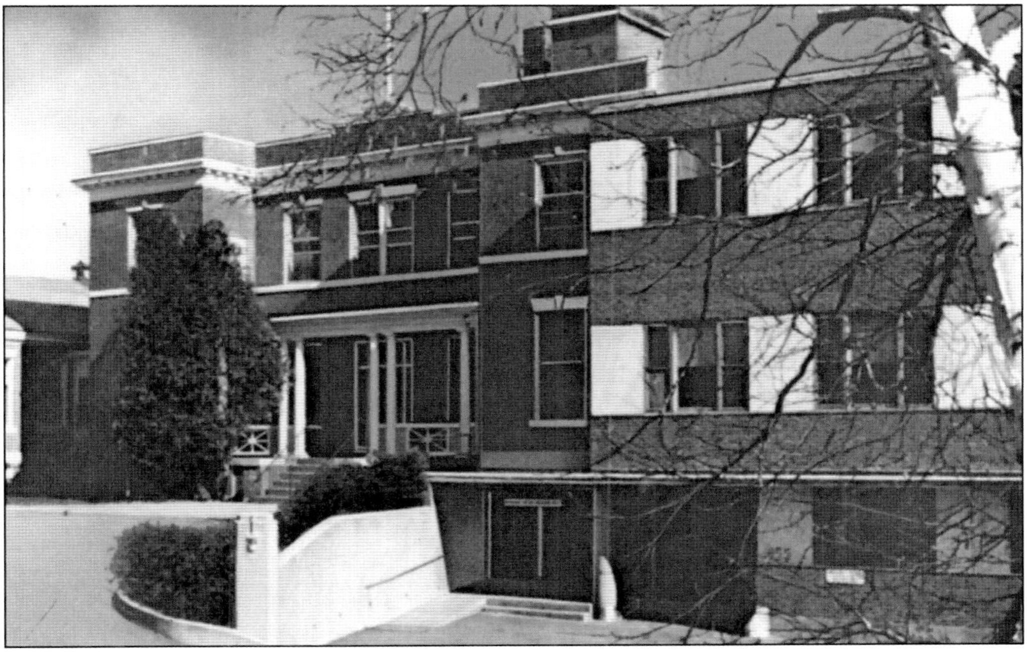

In 1953, another major building campaign was initiated, this time for a new two-story wing of the hospital. Two years later, in December, the cornerstone was laid, but it would be November 1957 before the grand opening. One of the highlights of the opening was the circus mural by Stuart Eldredge in the children's playroom. That restored mural can now be seen in the Children's Room at the Springfield Town Library. (HSP.)

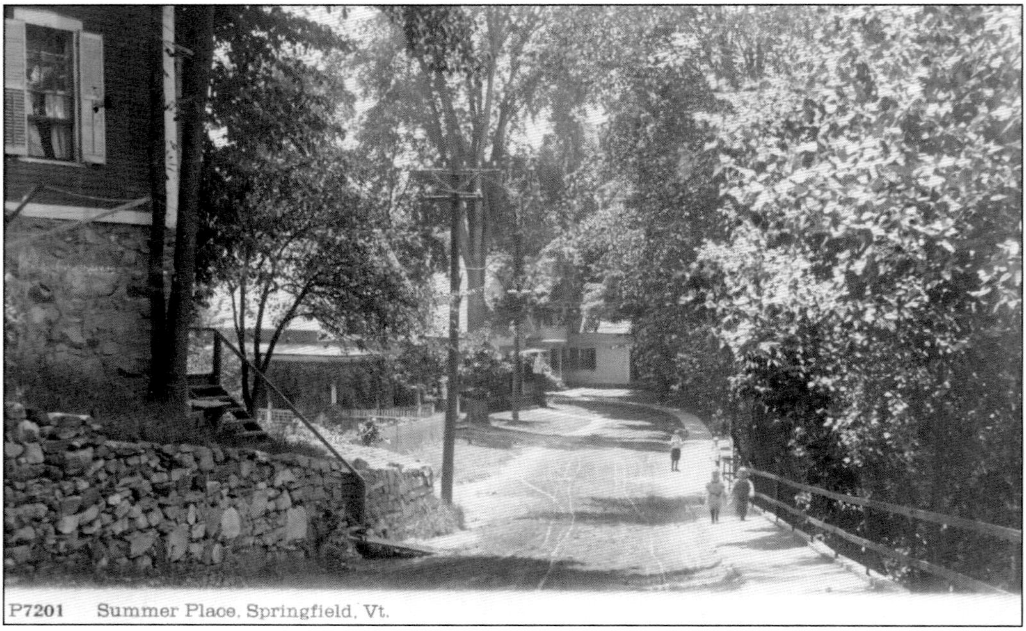


Now known as Wall Street, in 1889, this road only went to the first corner and was known as Summer Place. By 1905, Wall Street began at the end of Summer Place but only went as far as house No. 25. In time, the roads were combined and renamed Wall Street. It appears that this picture was taken before 1907, as the Brookline Apartments would be on the right after 1907. (HSP.)

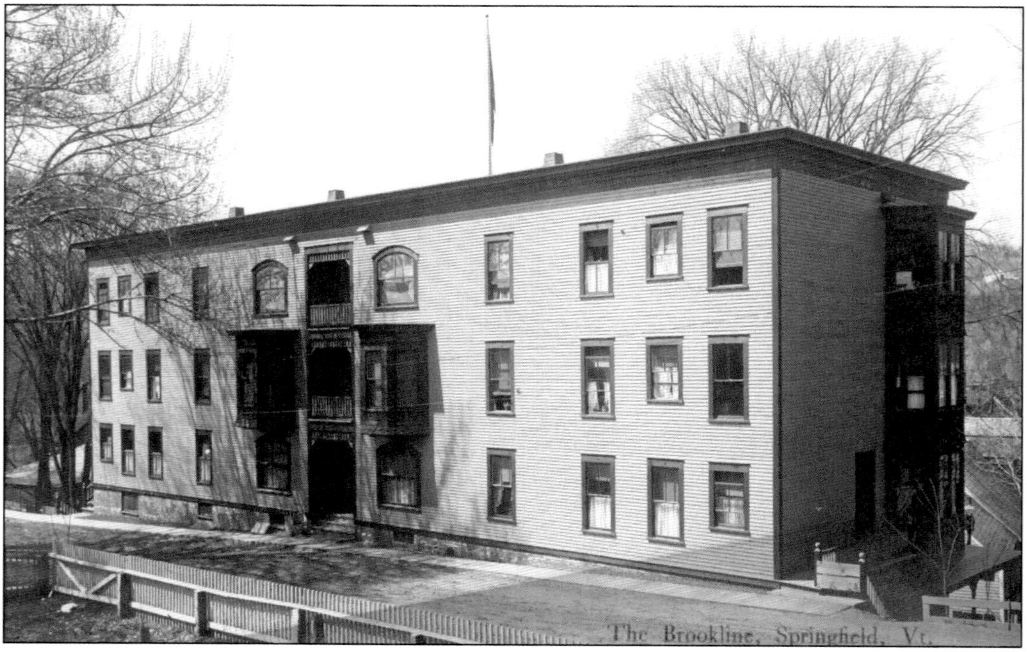


Brookline Apartments, erected in 1907, was the first building with rental units in Springfield. Perched on the edge of Wall Street, it had three stories above the road and three stories built into the bank. There was a separate furnace for each unit, which meant a long hike from the top floor to add coal. The apartments had gaslights, and each kitchen contained a coal stove. The building has survived for over 100 years. (STL.)

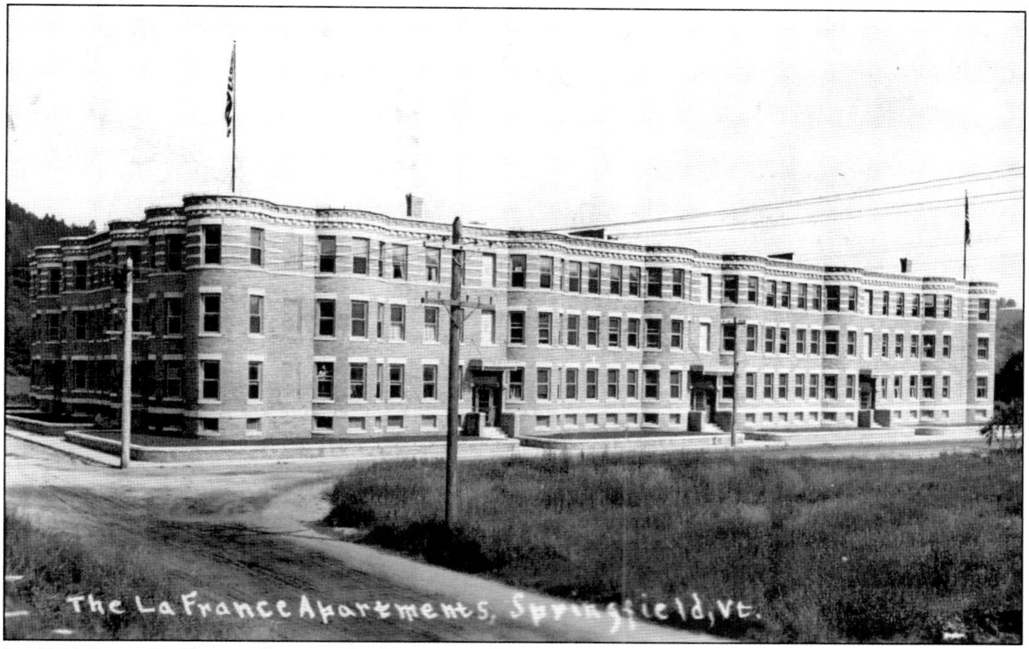


In 1916, Louis LaFrance from Holyoke, Massachusetts, built the 30-unit LaFrance Block at the far end of Wall Street. It remained the largest apartment block in town for over 60 years. Apparently, some of the two-bedroom units required one to go through either the living room or a bedroom and hall in order to get from the kitchen to the dining room. (HSP.)

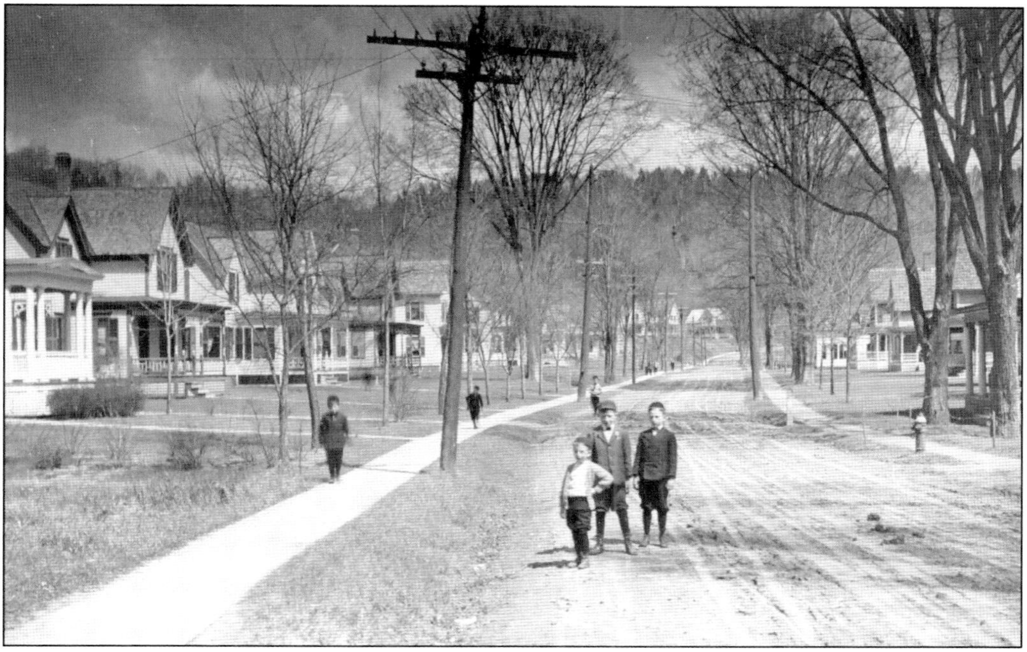

In this photograph of Summer Street, the boys look a little too dressed up for a ball game. There are other children coming up the street, so possibly they are all heading off to school. (STL.)

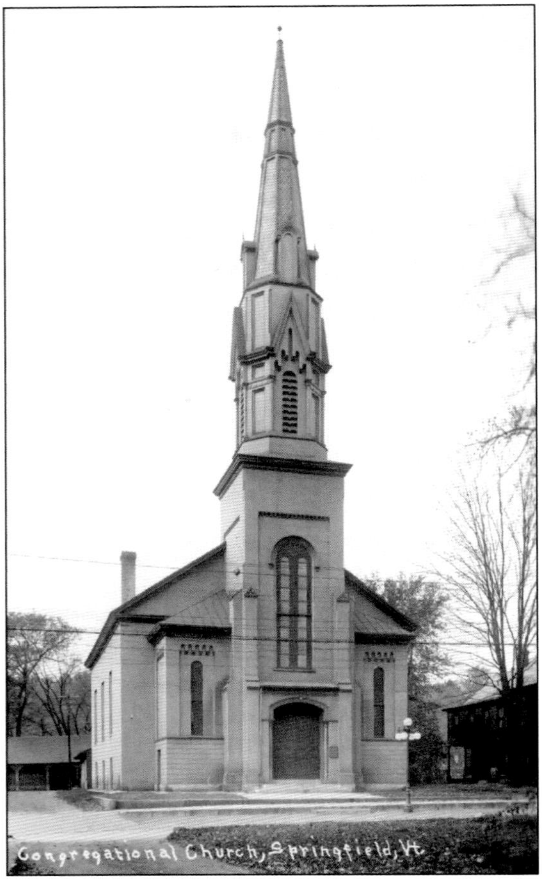


The First Congregational Church in Springfield was located on the common and no longer exists. The next church was built in 1833 and was located on Main Street. It was a fairly plain building with a small, simple steeple. In 1869, the facade was remodeled in the Gothic Revival style. The spire was extended, making it easy to locate in old photographs of the town. (HSP.)

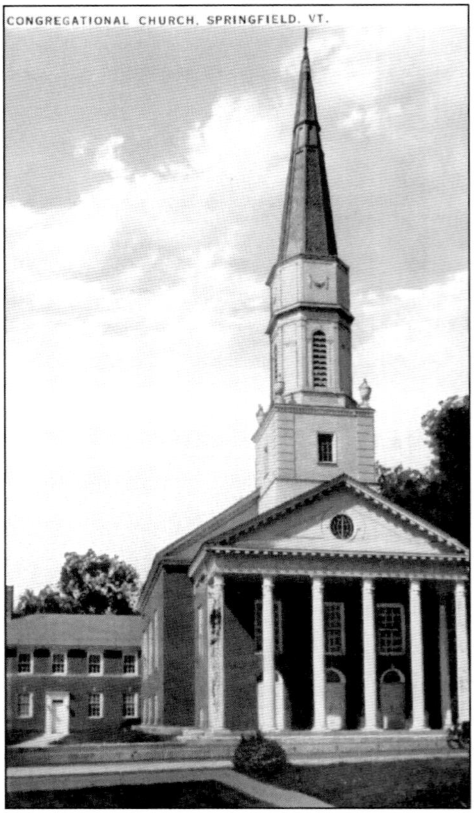


In 1929, a Colonial Revival remodel completely changed the church's appearance to what it is today. During this remodel, the steeple was lowered a bit and the portico added, giving the church a more traditional New England look. (STL.)

In 1843, the Methodists built this beautiful stone church just off the head of the square on Valley Street. The first clock was installed in 1844. It was an eight-day clock, but due to complications when the church was wired for electricity, it became necessary to wind it twice a week. In 1866, a bell weighing 1,410 pounds was purchased and first rung on December 12, 1868. (STL.)

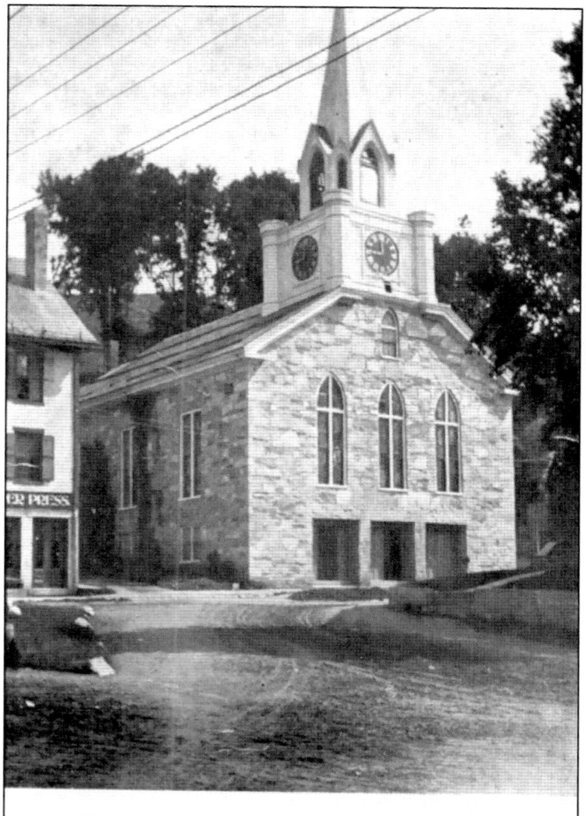


THE M. E. CHURCH, SPRINGFIELD, VERMONT.

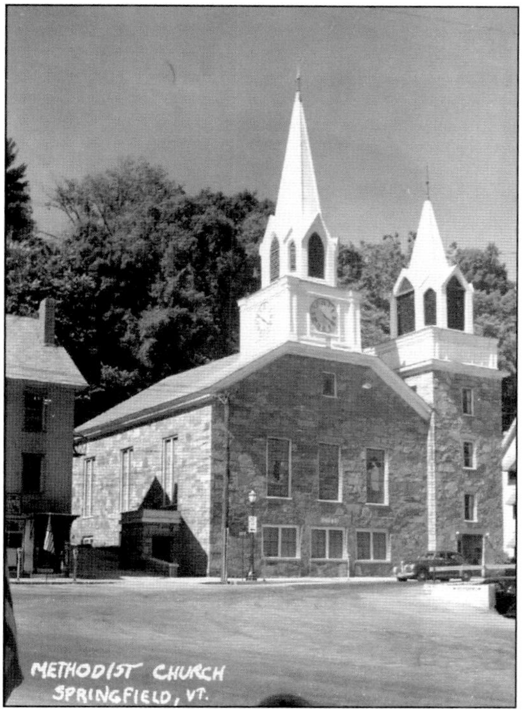


In 1917, an 18-by-22-foot tower with an attached 16-by-44-foot addition behind it was annexed to the right side of the Methodist Church. The entrance was moved, and the Gothic windows were replaced with rectangular ones. At the time of the dedication, there was no steeple on the tower. (STL.)

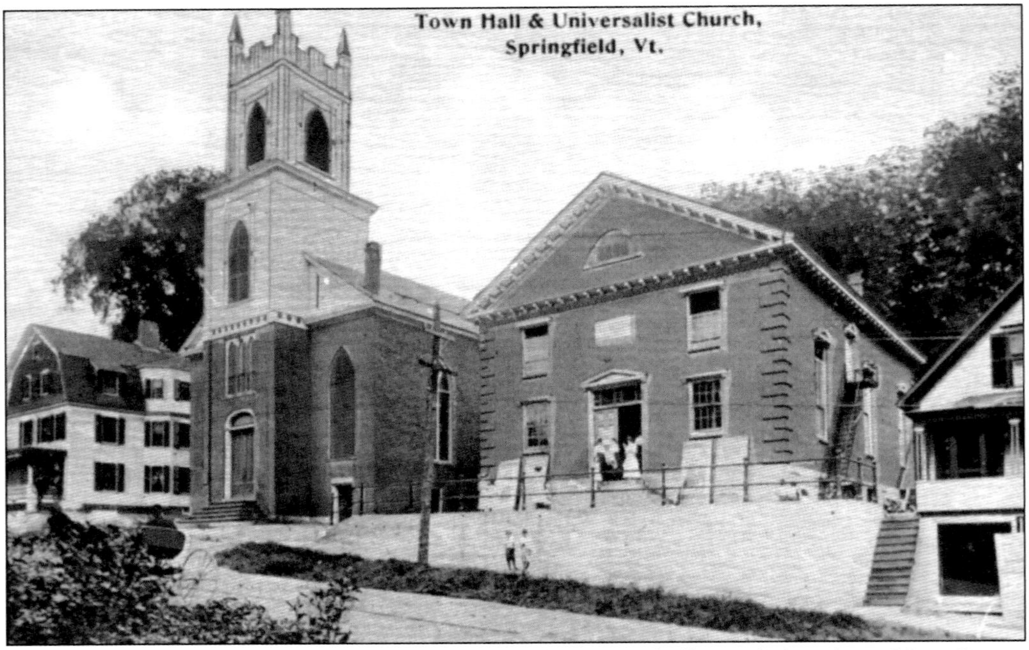


For close to 100 years, the Universalist Church and the town hall sat side by side on Main Street. There was hardly a photograph taken that did not include both structures. The church was built in 1833 and also served as the town meeting place until the town hall was completed. Today, the town hall stands alone, as the church was torn down in 1971. (STL.)

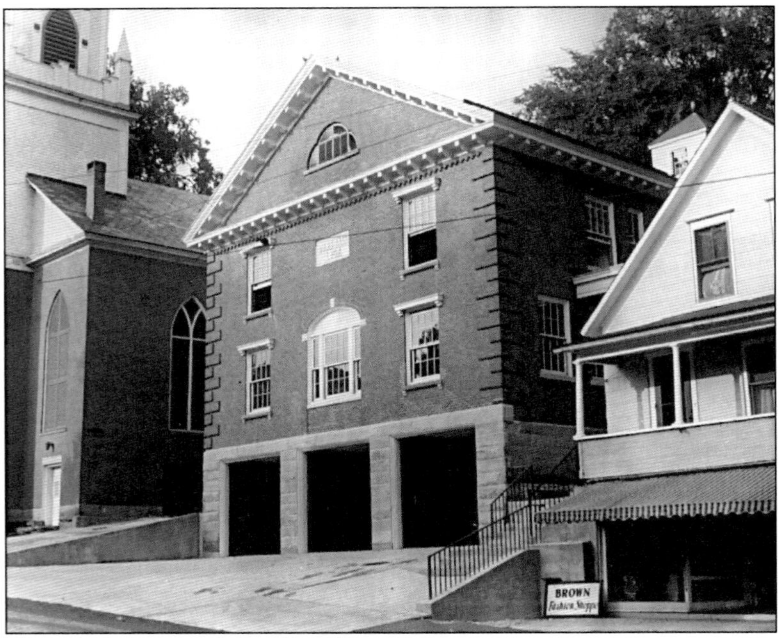


Town hall, the two-story, one-room social and civic center, was built in 1858. In 1939, it was turned into a three-story, multipurpose municipal building. This change included splitting the main room into two floors and renovating the basement to hold the fire department. (BL.)

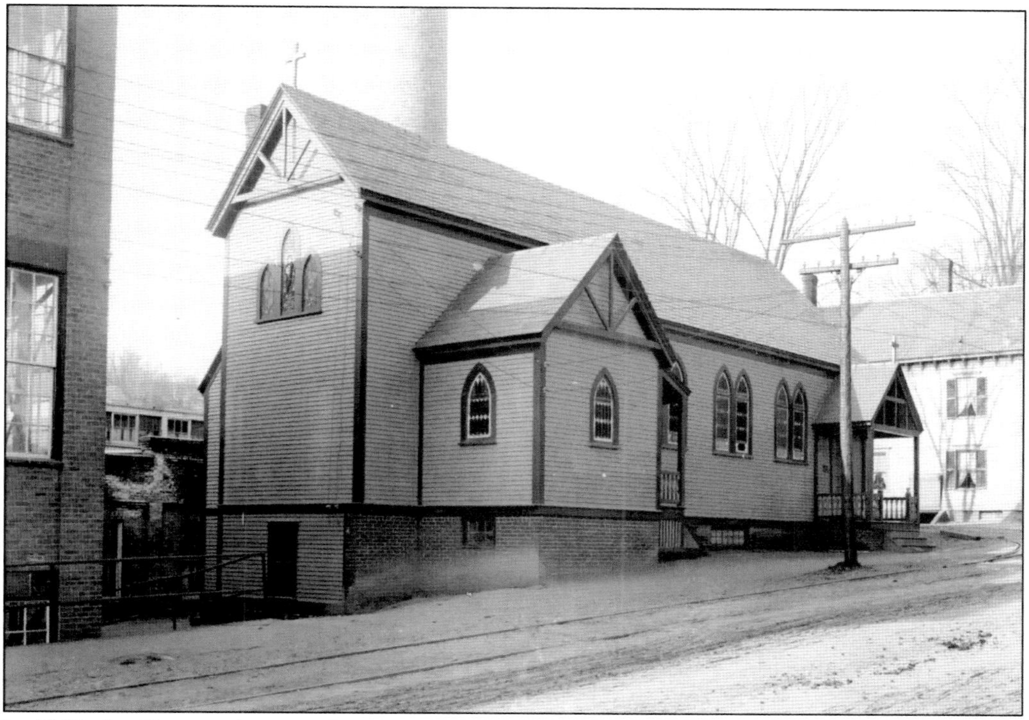

In 1900, the Episcopalians—who had been sharing meeting space with the Universalists—were finally able to erect their own church on Main Street next to the J&L building. This sufficed until 1955 when more space was needed. Dr. Jeremy Brown donated four acres of land on Fairgrounds Road, and the church was picked up and moved the mile and a half to the new site. It took two days to move the building. (Lovejoy Tool Company.)

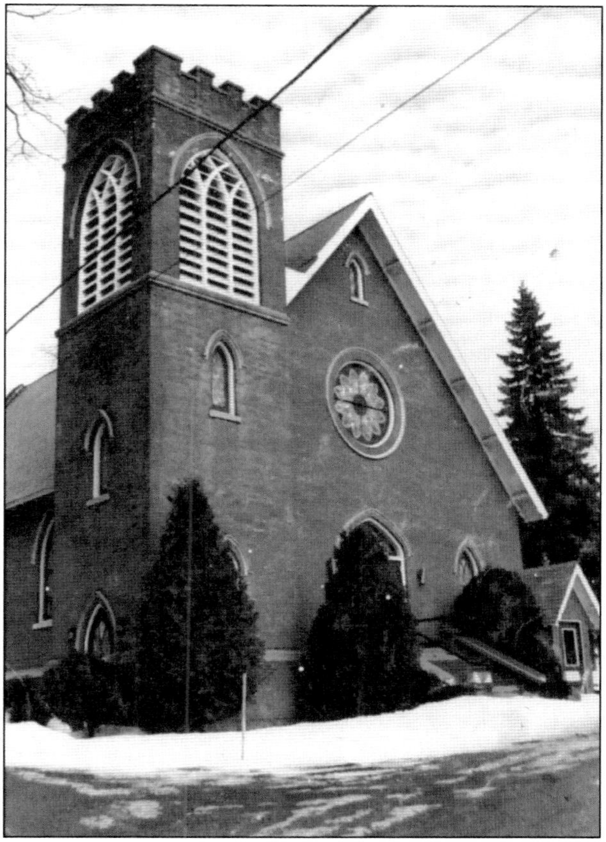

The last church built on Main Street was the Calvary Baptist. The church purchased the last two houses on the east side of the river next to the Clinton Street Bridge. They met in one of these houses while preparing to build their church. In October 1924, the cornerstone was laid, and the first prayer service in the new church was held in April 1925. (Arthur Bryant.)

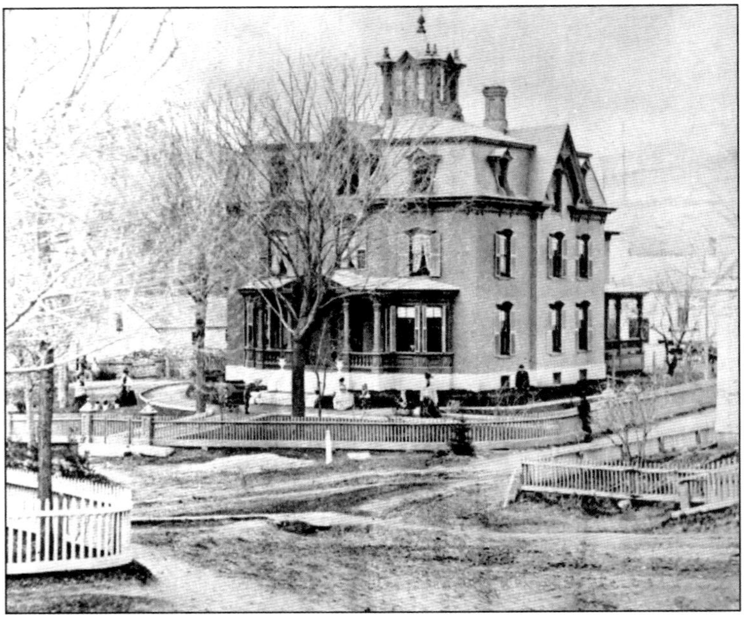

This landmark house, located on the corner of Summer Hill and Pleasant Streets, was built in 1869 or 1870 by Amasa Woolson. His son W.D. renovated the outside of the house, and in 1909, he built an impressive garage. The house suffered a substantial fire in 1920 and was uninhabitable. In 1925, the Catholic Church bought the property and renovated the house to be used as a rectory. (STL.)

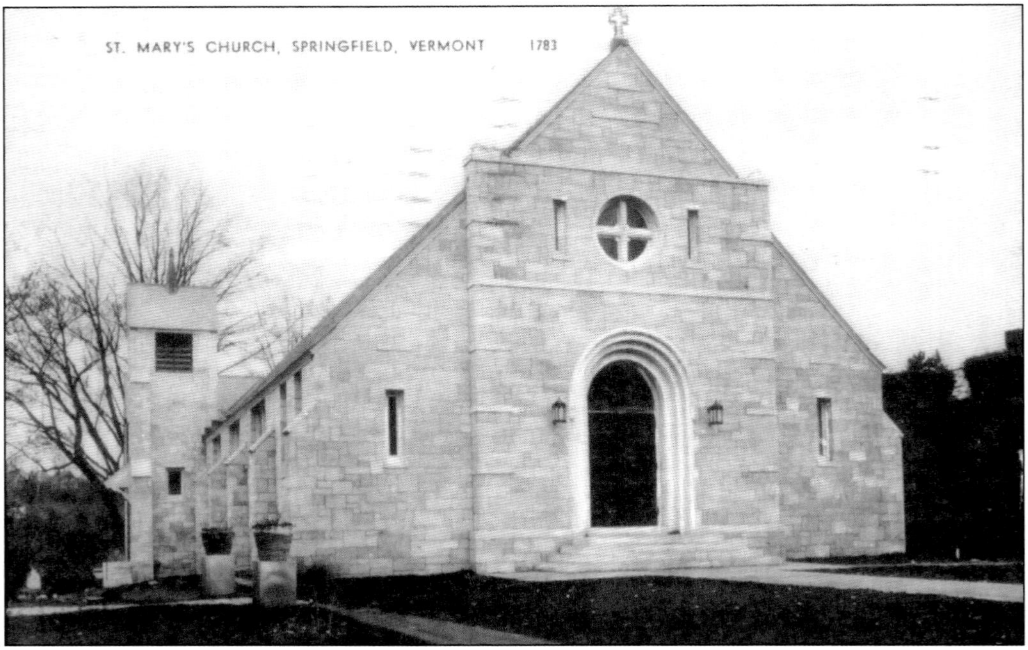


To replace the small wooden church on Mt. Vernon Street, a new one was built on Pleasant Street between the rectory and the garage. Started in April 1930, it was built of white Vermont marble in the Romanesque style and was dedicated in November. The oak interior was described as having a tapestry facade of brick with oak battering and a marble altar. (STL.)

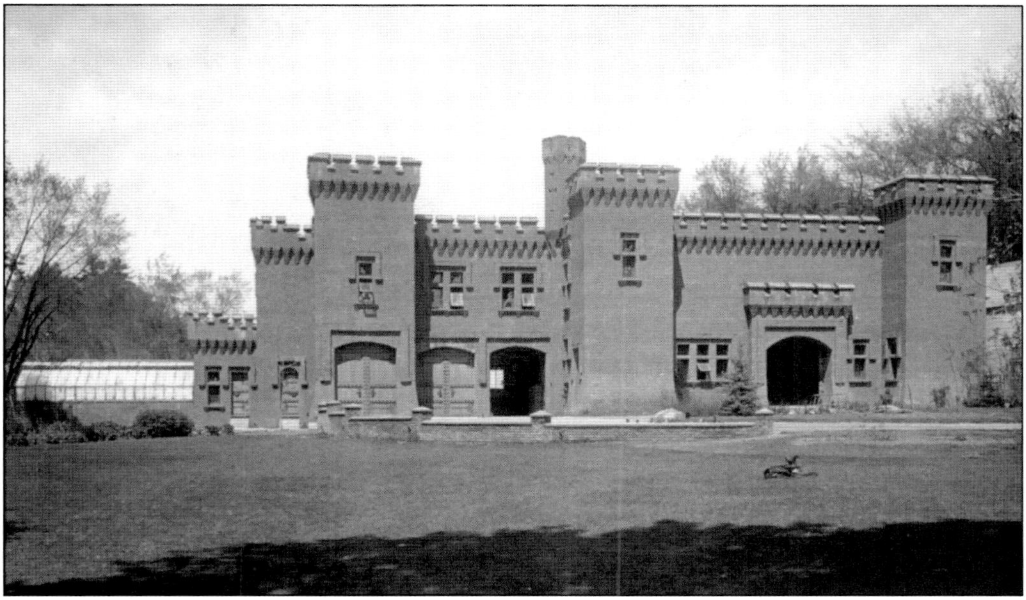

Although the church was built in 1930, it was 1951 before the garage was converted into a parish hall. Originally having stalls for five horses and room for several carriages and four or five cars, the first floor garages were retained, but the second floor renovation included an auditorium, kitchen, and other facilities. The building was torn down in 1969 and replaced with the Nolin-Murray Center. (MAC.)

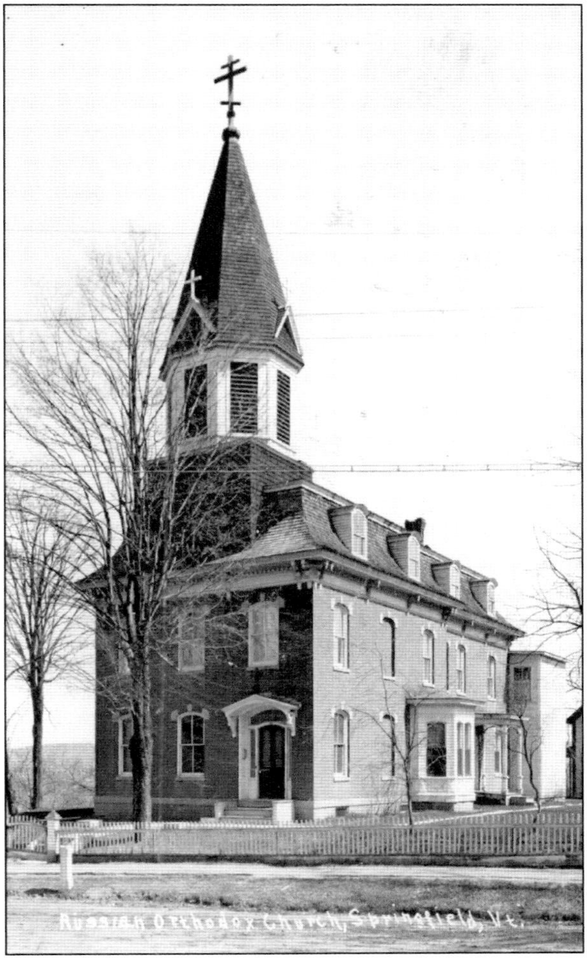


In the 1890s, Springfield became a different town when the first Russian and Polish immigrants arrived. Initially coming to work on farms, these immigrants soon gravitated to the shoddy mill and then to other companies. The Russian Orthodox Church was a converted private home. The new steeple was dedicated on October 19, 1911, at 5:00 a.m. The dedication was held at such an early hour in order to avoid interfering with the parishioners' employment. (HSP.)

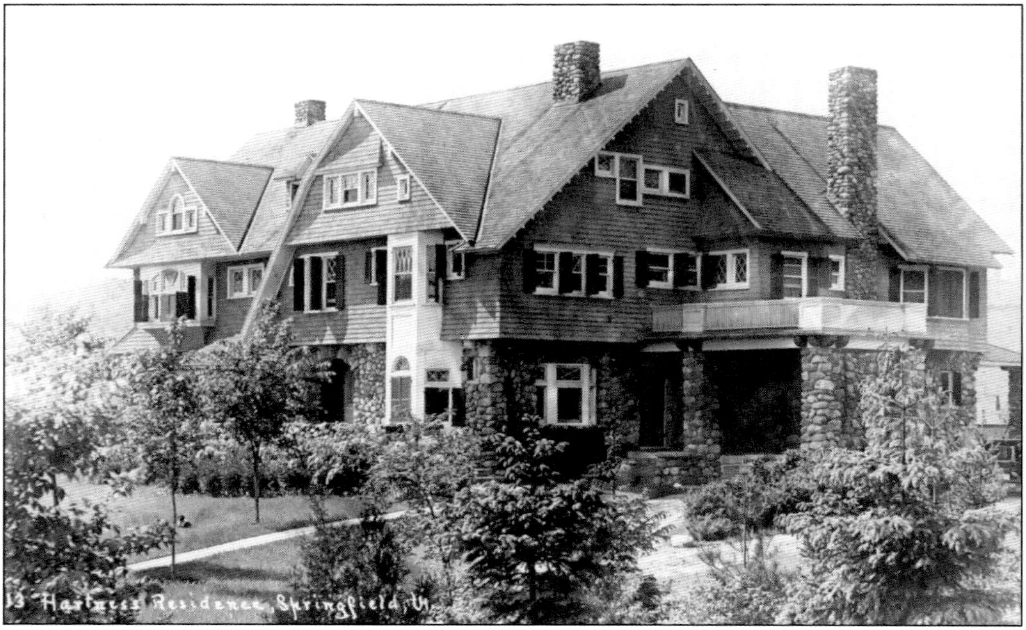


James Hartness built this home in 1904 and lived there the remainder of his life. He resided here while serving as governor of the state, and he entertained Charles Lindbergh here. As his residence, the main entrance was through the porte cochere on the right. This was enclosed and the space turned into the dining room when the house was remodeled in 1940 to become the Hartness House Inn. (STL.)

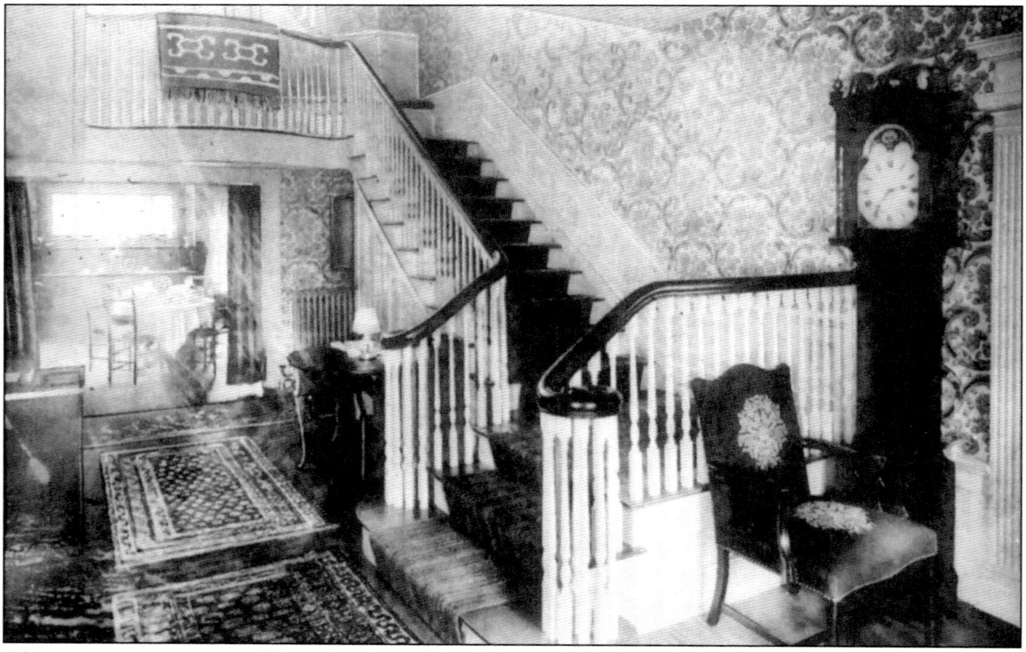

There are not many public photographs of the interior of the house when James and Lena Hartness lived there. However, as the Hartness House Inn, this staircase has been featured in many wedding photographs. (HSP.)

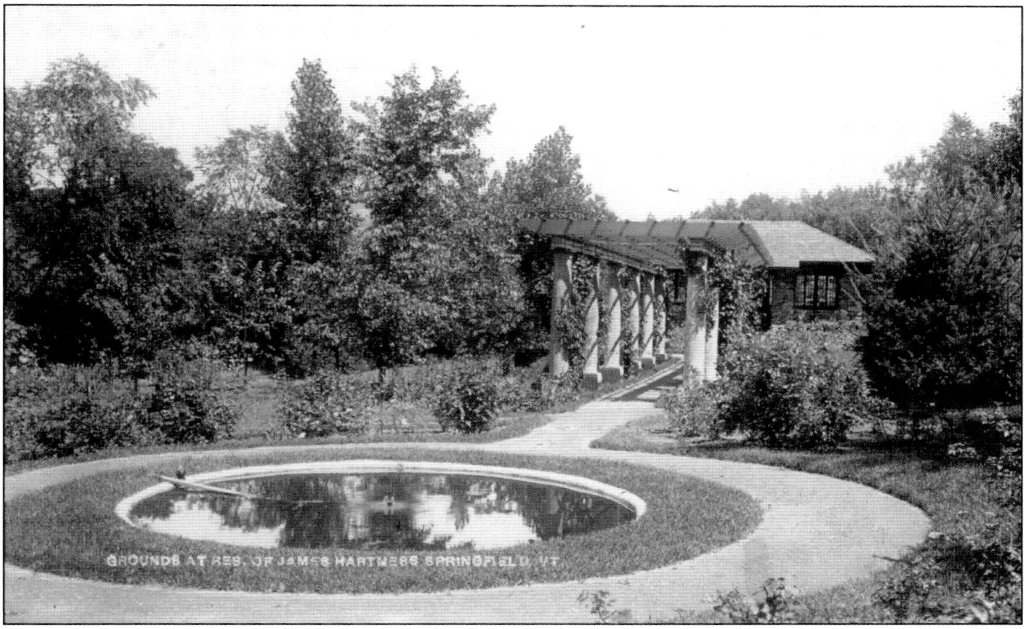


The reflecting pool, arbor, and teahouse (just to the right of the arbor) were all typical features of the formal gardens of the early 20th century. The Hartness House Inn now features the teahouse surrounded by a wedding garden and fountain. (HSP.)

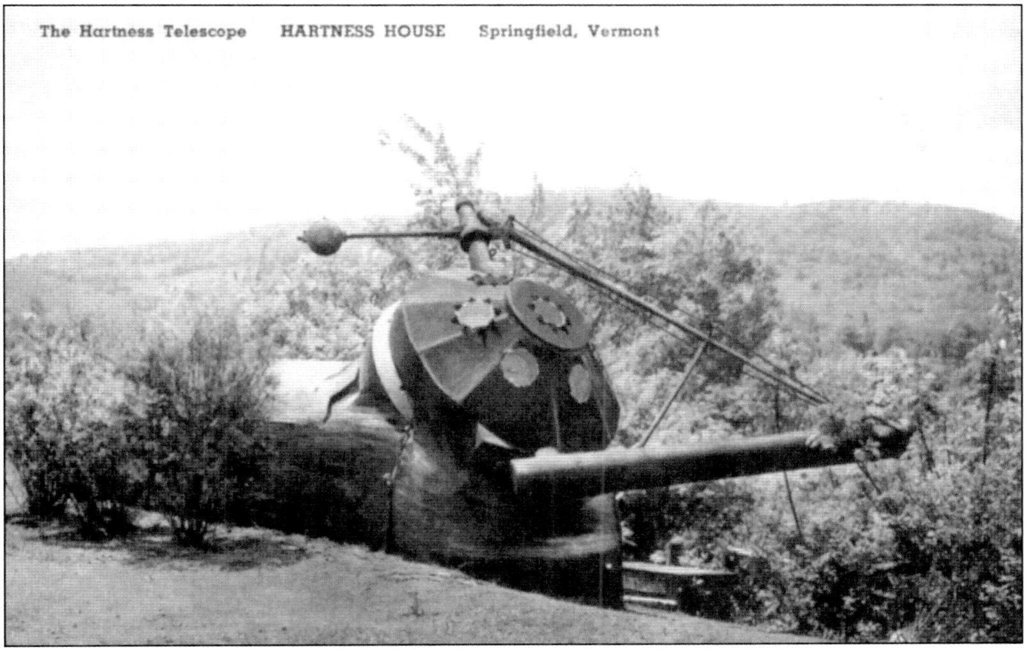


Hartness was fascinated with astronomy, and in 1910, he designed and built the Hartness Equatorial Turret Telescope on his front lawn. In 1921, in order to avoid the cold, snowy, winter nights, he completed an underground tunnel leading from the house to the telescope. The tunnel had several rooms and the added advantage of giving him a quiet place, away from the noise of family life, to work. (HSP.)

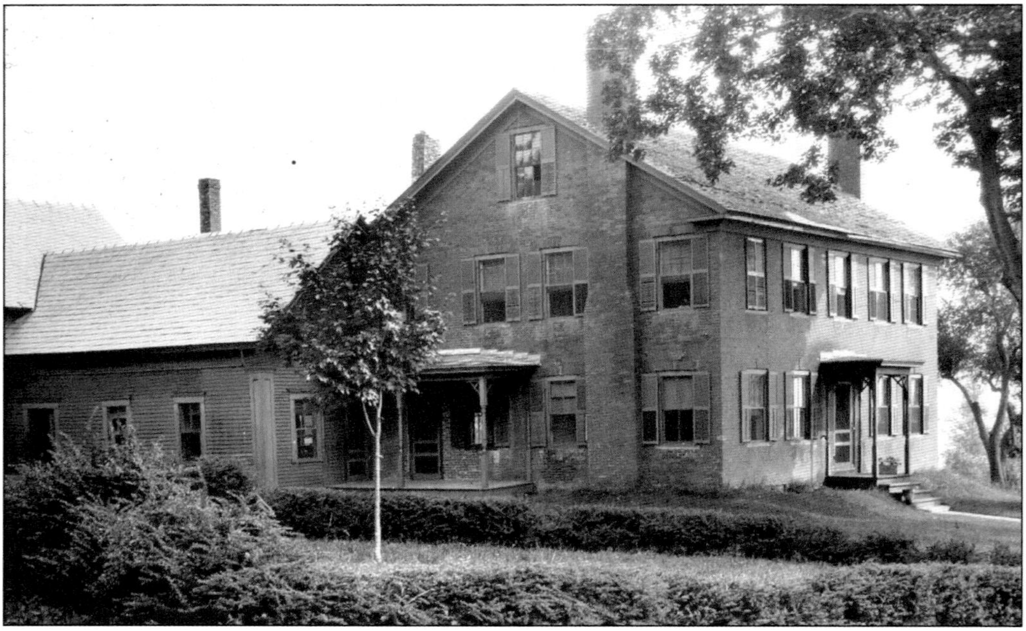

It is estimated that Smiley Manse was built in 1814 by Robinson Smiley, the pastor of the Congregational Church and the first settled preacher in Springfield. Located on Cherry Hill, it does not seem so far out of town when one realizes that, at the time, the church was on the common. When Sen. Ralph Flanders bought it, there were eight rooms, six fireplaces, and no central heat. (WGS.)

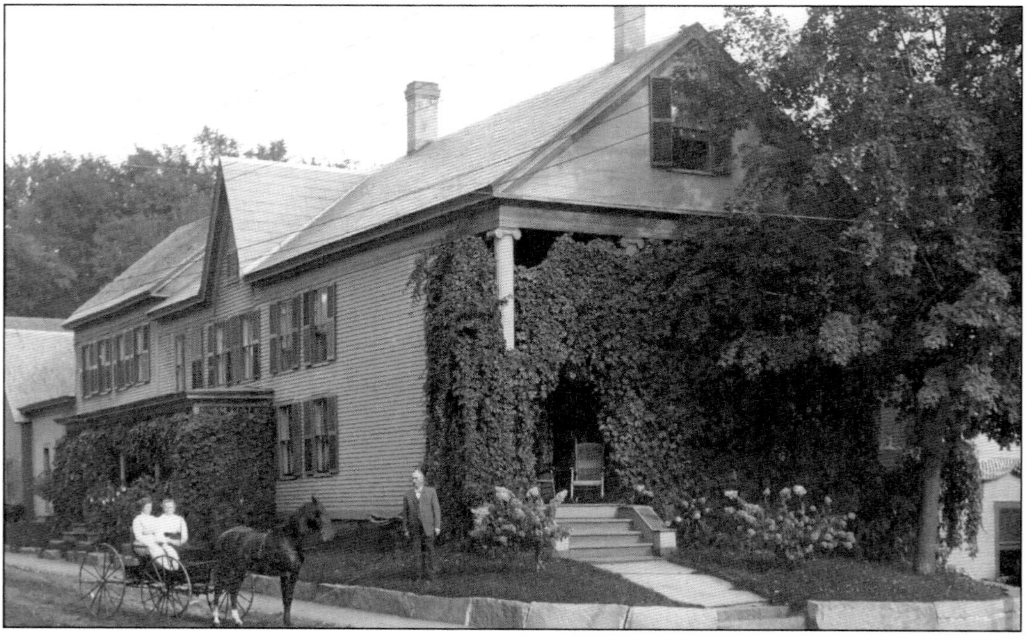

This house was located right across from the Methodist Church. One can always recognize the columns in old photographs, but it was not always pretty. In the earliest photographs, the porch steps are right on the street. This photograph was taken in the 1890s, when Frederick Ellison owned the building. Later, it would be used for stores and offices before finally being torn down in the mid-1900s. (STL.)

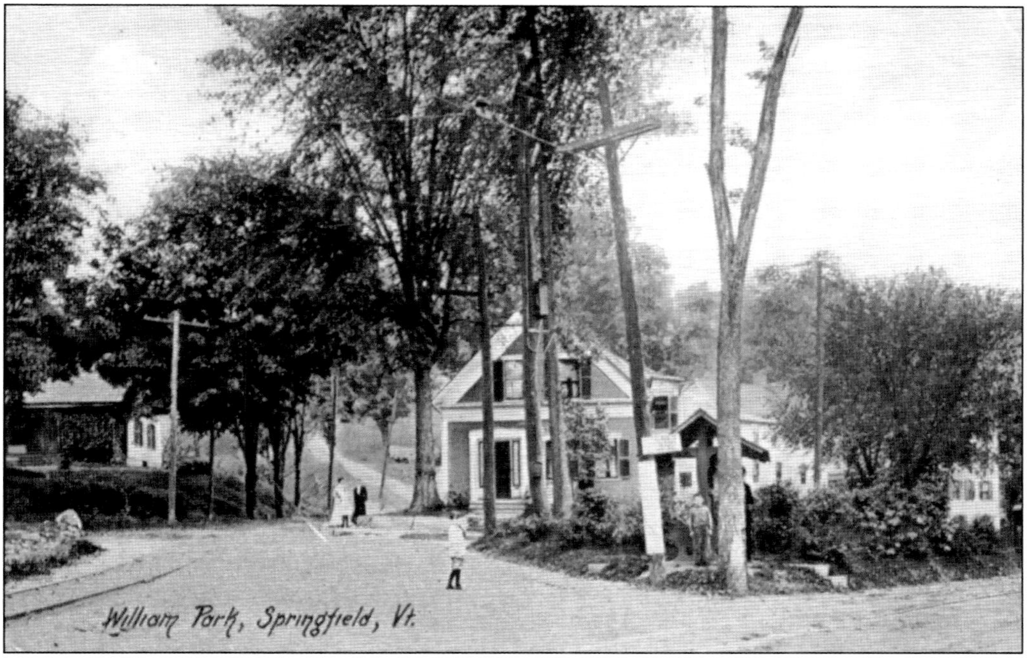


This is a view of Williams Park where the trolley tracks split for Mineral and Main Streets. The image can be dated to after 1903 because the bottom of the telephone post is painted white, which indicated a trolley stop. The house in the middle is where the Whitcomb Building is now located. (STL.)

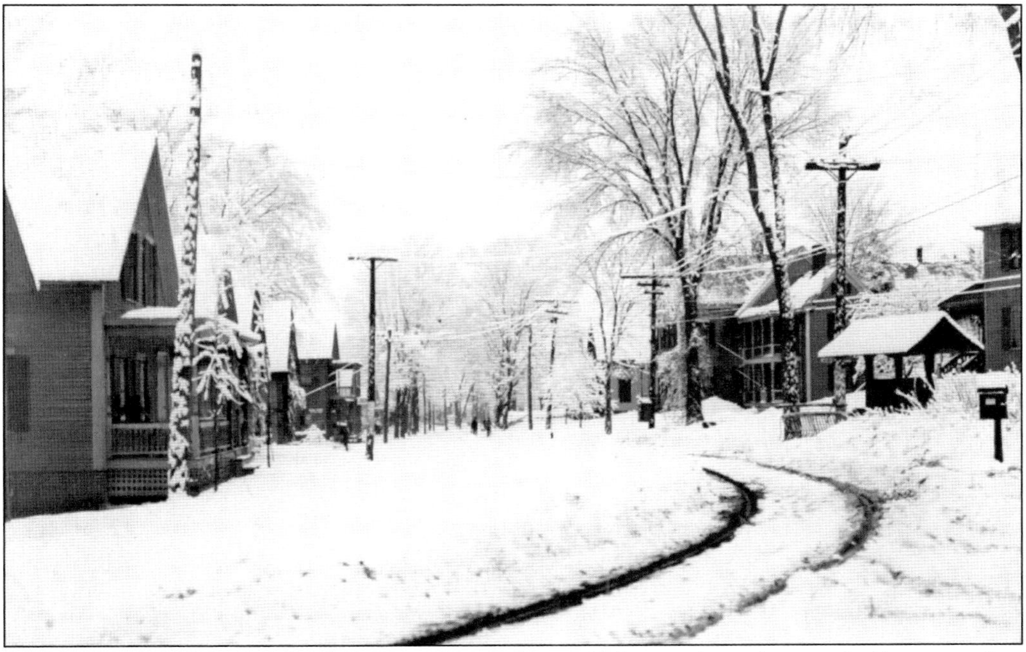

This is a winter photograph of Clinton Street after 1897. It must have been a challenge to plow the trolley tracks but still keep the streets safe for horses and sleighs. Sleighs were known to tip over if they slid into a plowed track. (HSP.)

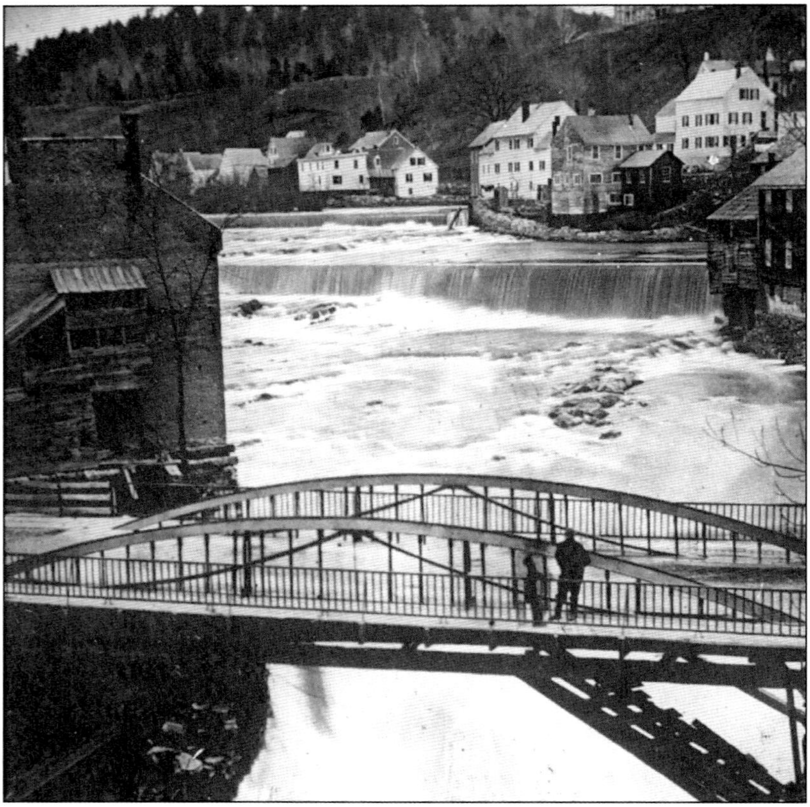

Comtu Falls is Springfield's landmark. The first bridges were made of four huge hemlock trees laid across the falls. One log was replaced every four years. Around 1869, the town decided to replace the wooden bridge with this iron one. Notice that the sides of this bridge go directly into the flooring. The flood of 1869 took this bridge out shortly after it was installed. (WGS.)

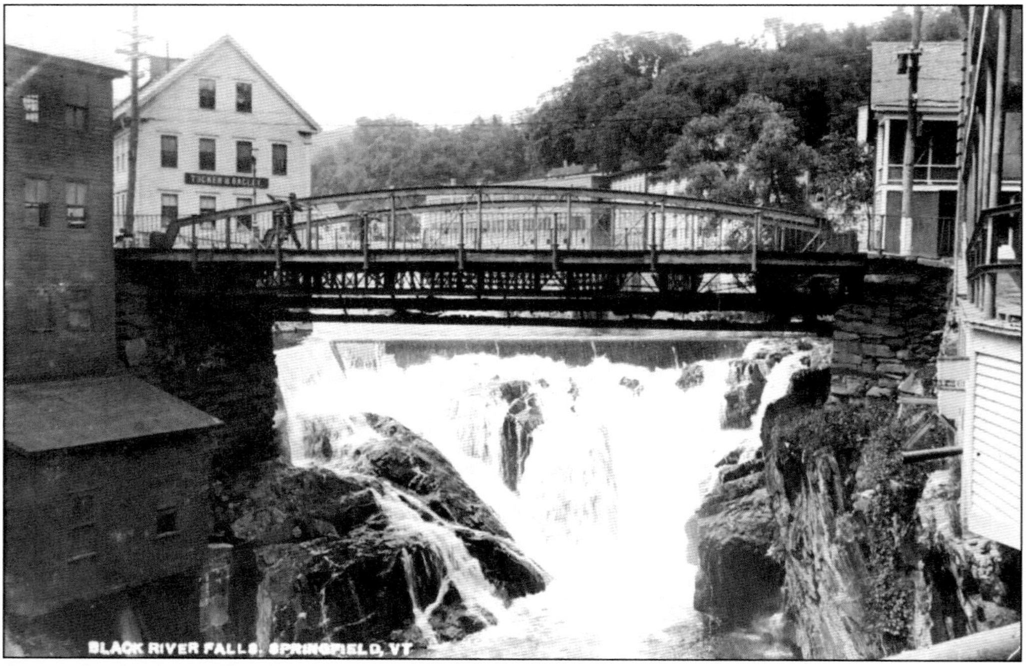


A wooden bridge sufficed until the second iron one was installed. This bridge can be identified by the clipped ends of the sides. It stood until 1916 when it was deemed unsafe and replaced with the cement bridge that currently crosses the falls. (See chapter 10, page 123.) (HSP.)

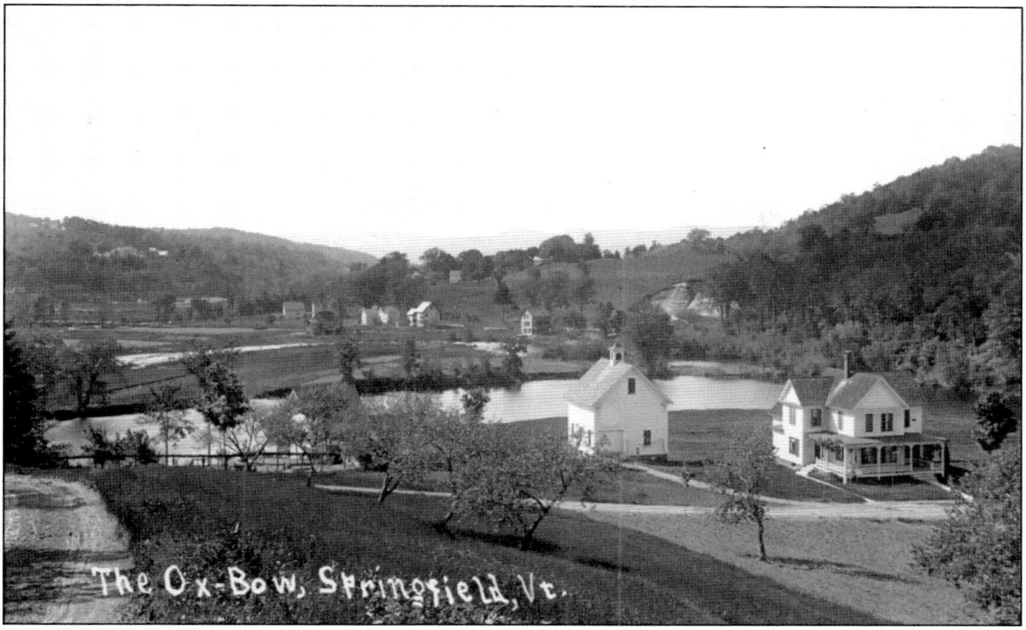


This photograph of the Oxbow was taken from Prospect Street. Before the town pools were built, there were two swimming areas in the river. One was to the left of this photograph in the bow, and another was by Riverside. After the flood control dam was built in North Springfield, this area was developed into the Springfield Shopping Plaza. (STL.)

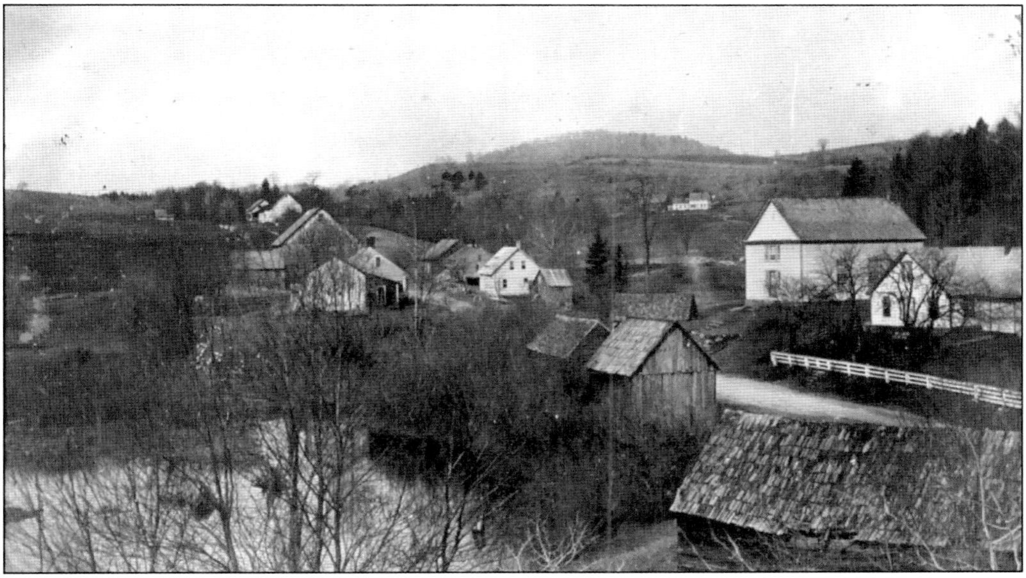

The original Chester Road (before it was redesigned and moved) is seen in this photograph of West Springfield. The large building on the right is the Reformed Methodist Church. In 1885, the church was reorganized; a lease was given to the board of trustees for as long as it was used as a church. The church and the pond are gone, and the thoroughfare is now called Cutler Road. (WGS.)

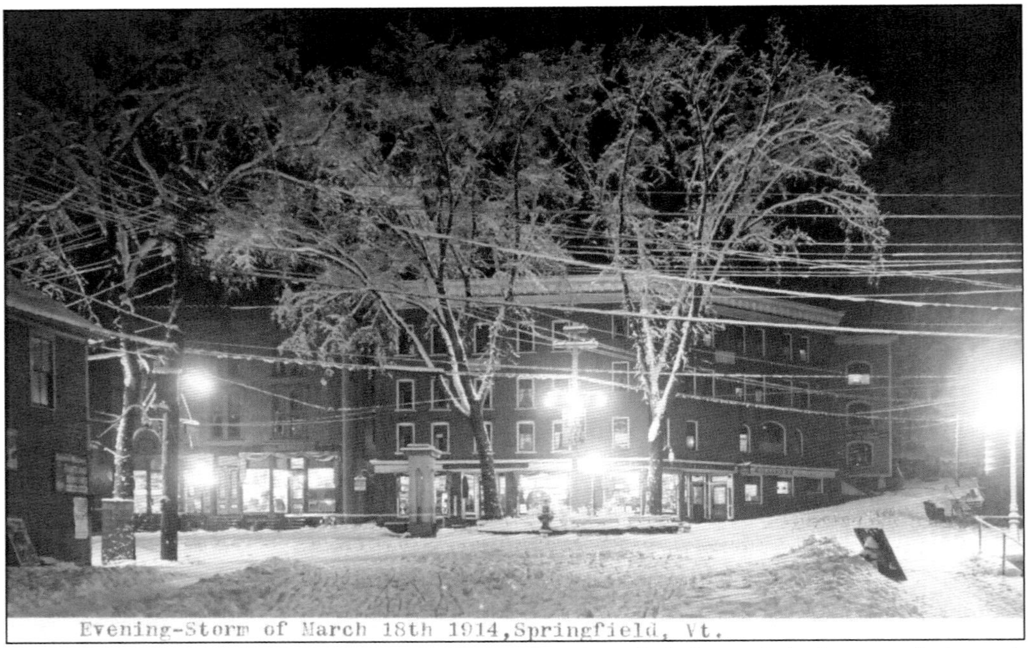


This pretty photograph was taken on March 18, 1914, from Park Street. It looks towards the park in the square and the Lincoln/McKinley Block. (STL.)

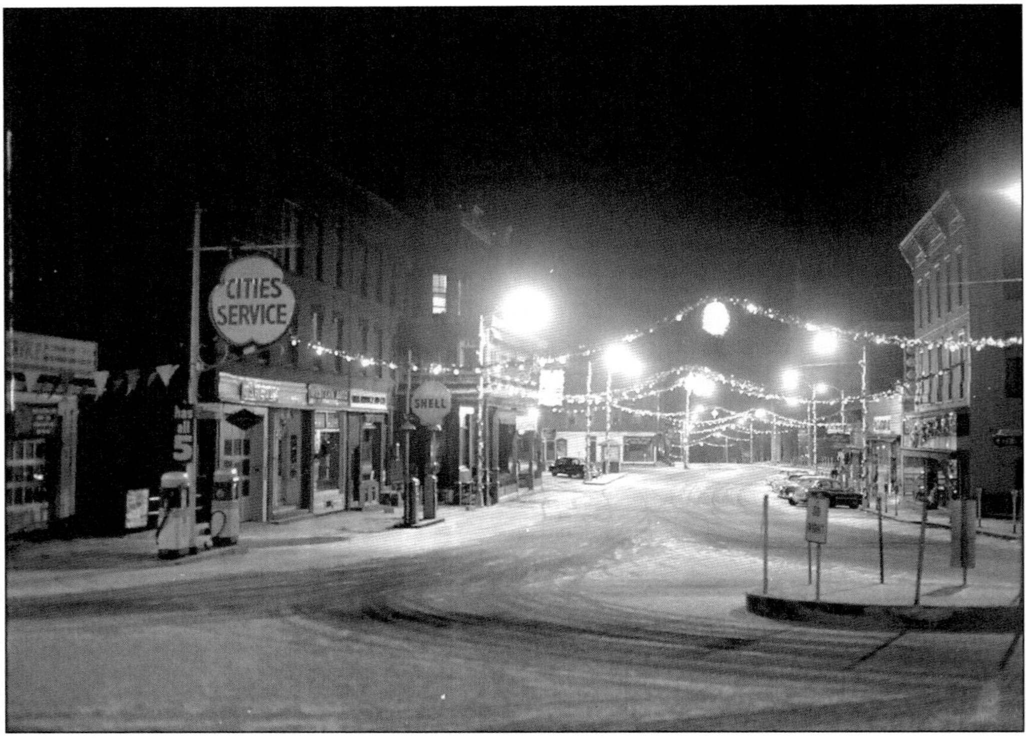


In this image, Springfield looks like a quintessential 1950s small town decorated for the holidays. Colored lights are strung over Main Street, and stores are lit up. Life seemed simpler. One shopped on Main Street or ordered from the Sears & Roebuck or Montgomery Ward catalogs. Most toys needed a child, not a battery, to make them run. (BL.)

# *Four*

# EARLY BUSINESSES

Once the dams had been built and waterpower harnessed, mills came and went (generally destroyed by fire or flood) along the banks of the Black River. Around 1774, there was a sawmill on the first dam in Goulds Mills. At about the same time, William Lockwood built a mill at the big falls farther up river. He also laid two large hemlocks across the falls, connecting the two sides of the river. By 1795, Lester Fling had built the first gristmill above the big falls on the east side of the river.

As the swampy land was drained, a town began to take shape. Isaac Fisher's cotton factory was built in 1811, and Col. Jonathan Williams's woolen and cotton factory was established in 1812. The creation of small manufacturing businesses and the new road from the Cheshire Bridge to the falls could be considered responsible for the birth of Springfield, a town that would eventually send the products of its skilled and inventive citizens to all parts of the world.

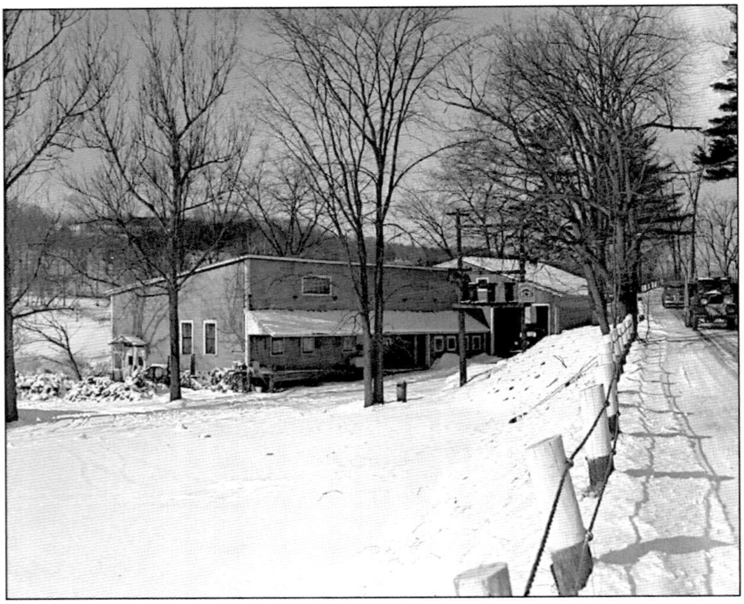

Ice was harvested from the Black River (near Riverside) until it silted up after the 1927 flood, at which point ice was brought in from elsewhere. In 1929, Clare Parker, owner of the icehouse (corner of Routes 11 and 106), installed ice-making equipment. He could produce 10 to 12 tons of ice per day. As electricity became available in the outlying areas, the icehouse eventually went out of business in 1953. (BL.)

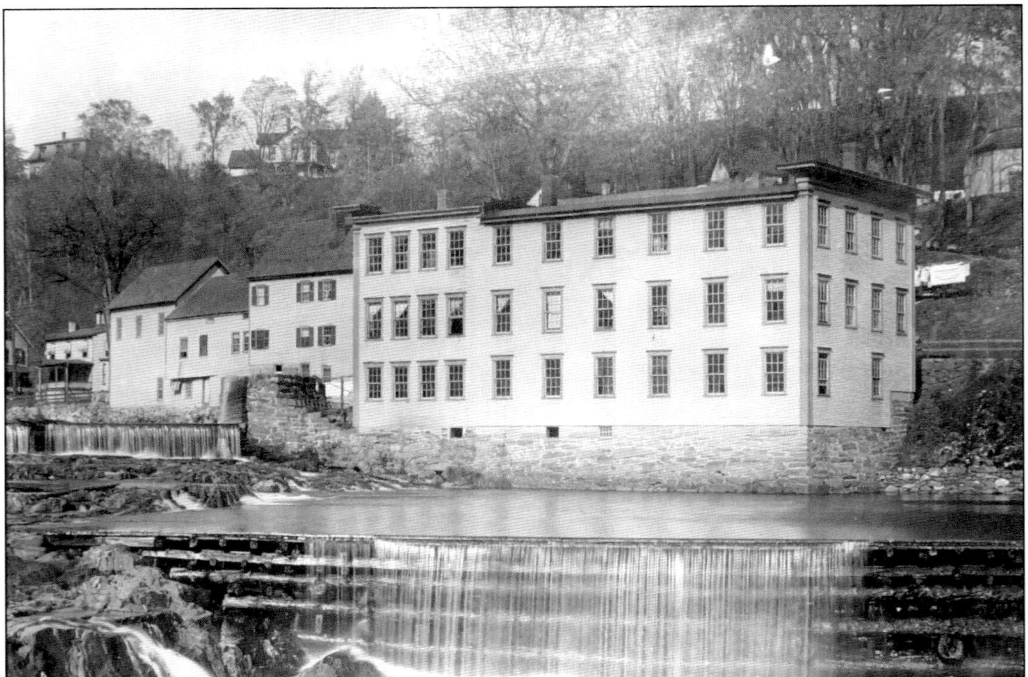

Farley B. Gilman came to Springfield in 1847. He established the Gilman Shop in 1854 and constructed this building in 1869. He invented several different styles of lathes for turning irregular forms, most notably hat blocks and shoe trees. He held many patents, and by the turn of the century, his machines were used throughout Europe. In 1892, his son Wilbert joined the company. (STL.)

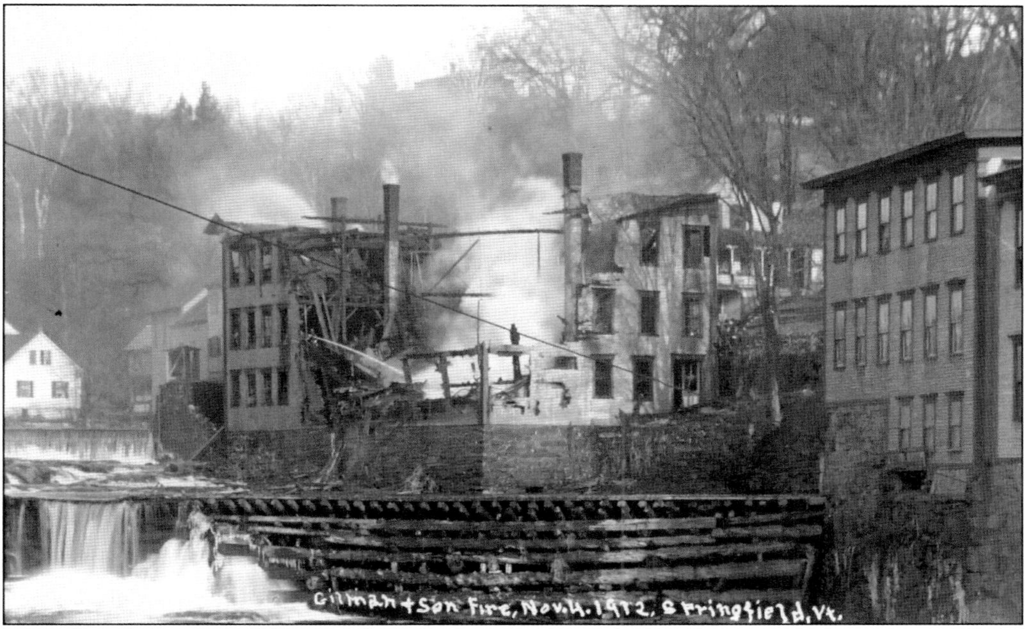


W.W. Slack bought Gilman & Son in 1909. On November 4, 1912, the plant was destroyed by fire. The fire department used six streams of water on the building. The office was saved, but the plant was a total loss. The 40 workers were told that the plant would be rebuilt. (STL.)

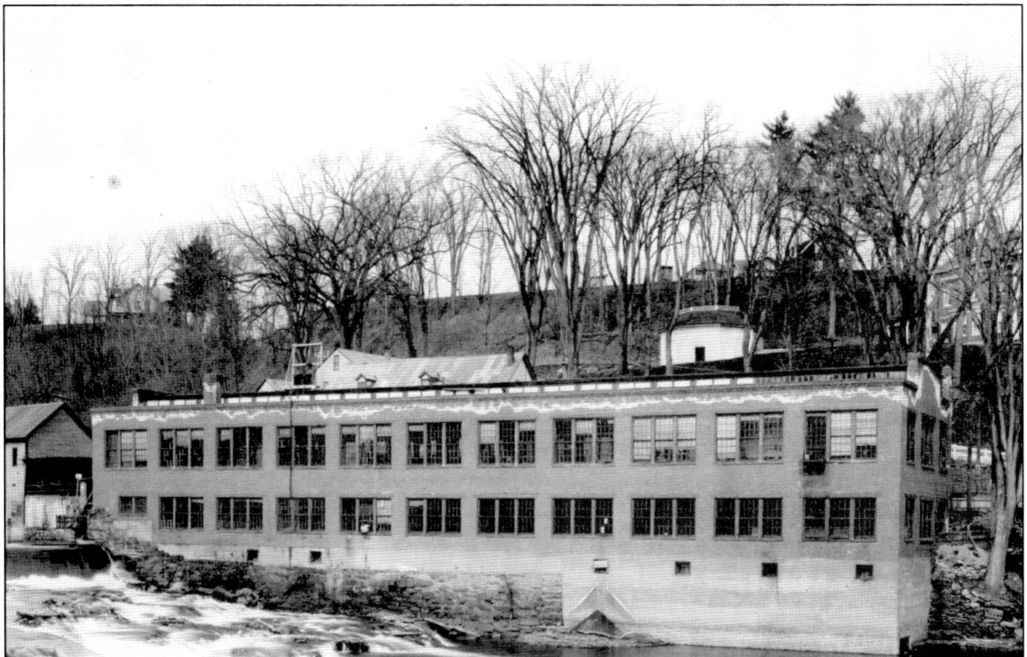

The company reopened in April 1913 in this new brick building. In 1916, controlling interest was sold to what would become United Shoe Machinery. Slack passed away in 1931, and the company assets were auctioned off in 1933. Unfortunately, the building suffered another fire in the 1980s. (STL.)

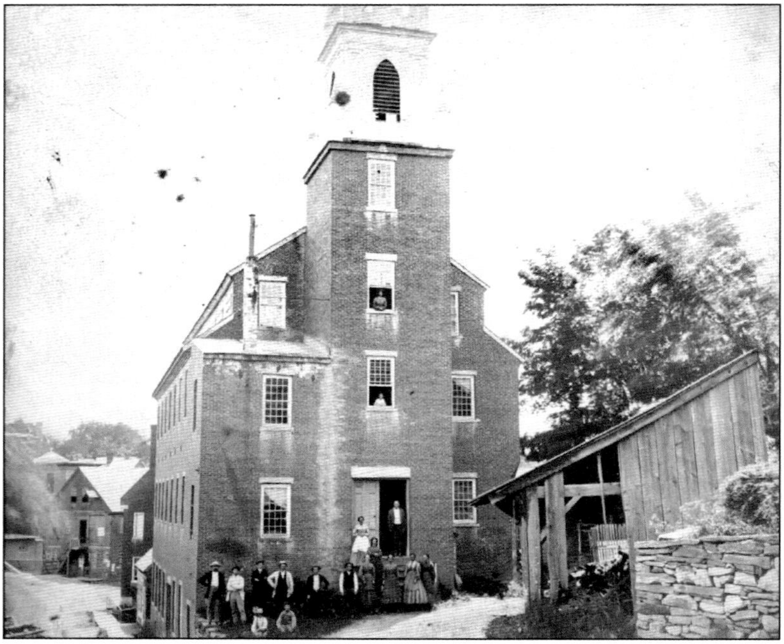

This brick factory, built in 1836, was the John C. Holmes and Company's cotton mill. It was located on the west side of lower Park Street. In 1894, this company installed telephones, making it one of the first to do so. In 1889, it was wired for electricity. As long as the bell tower is distinguished from church steeples, it is easy to identify the factory in old photographs. (STL.)

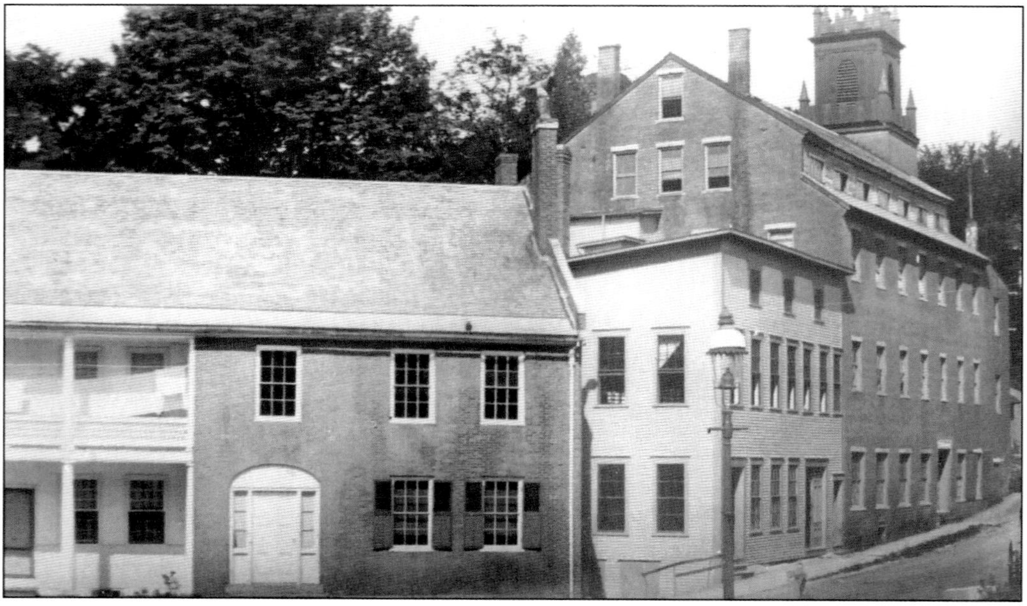

The Cotton Mill lost its third floor to fire around the turn of the 20th century and went into receivership in 1905. At that time, it was taken over and became part of the shoddy mill complex. It is now mostly used as office space. The building on the left, directly across from the end of the Falls Bridge, was a tenement and was later used for offices. It is now a parking lot. (STL.)

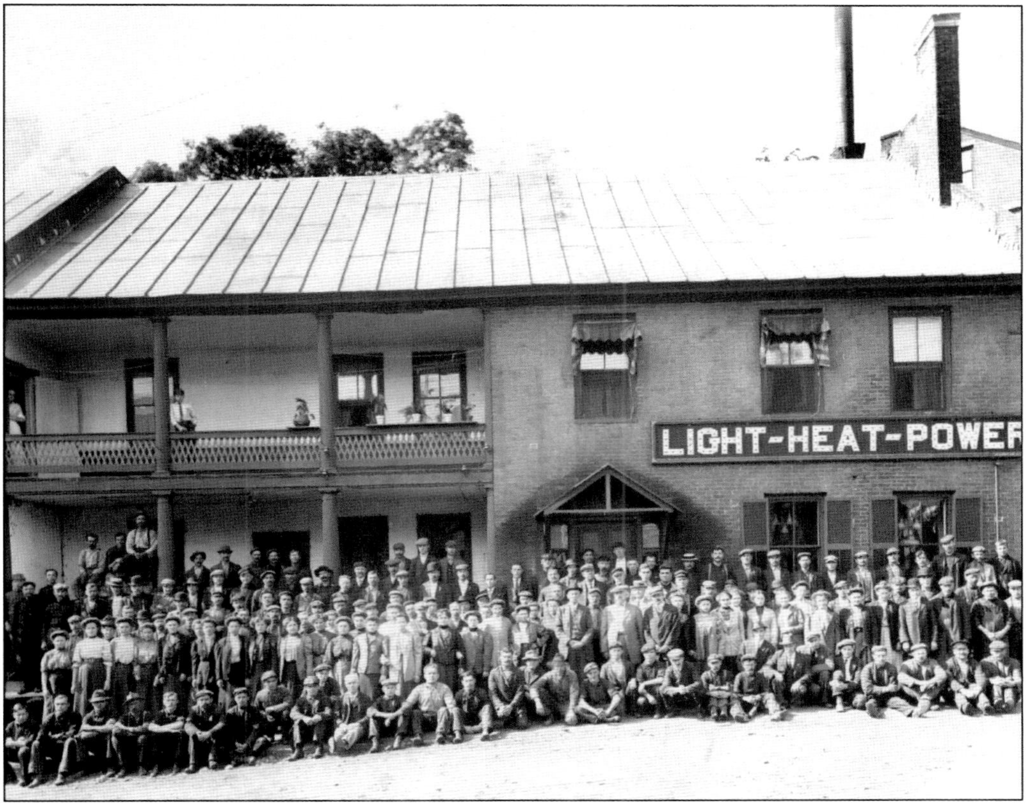


In 1911, the John T. Slack Corporation acquired the property, franchise, and other rights of the Springfield Electric Company. They built a new power station around the old one, and for two years they were the local electric public utility. In 1913, they sold out to Colonial Power and Light Company. This image may include all of the shoddy mill workers, as there are over 200 people shown. (STL.)

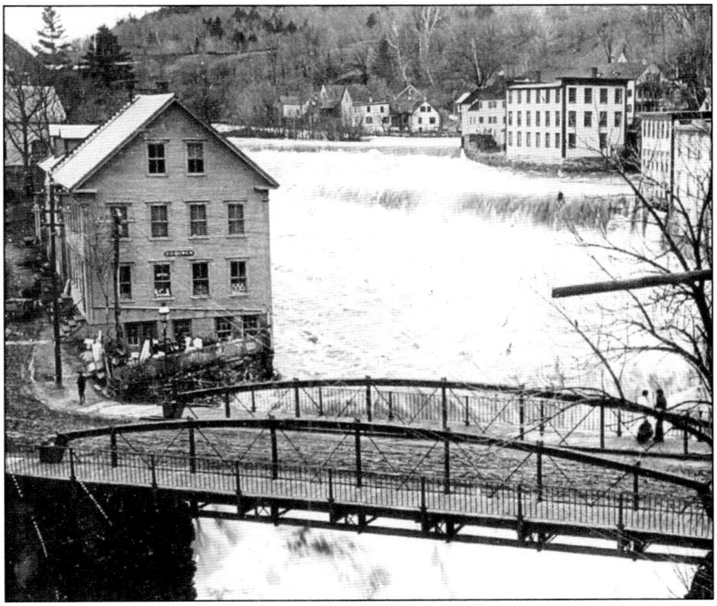

Moses H. Grinnell came to Springfield in 1841. He invented the Grinnell process of polishing marble by rubbing two slabs together with sand and water between them. After several owners, Franklin Barney bought the business (white building on left) in 1861. It acquired an addition on the front and went on to house a grocery store, flower shop, restaurant (the Purple Cow), dress shop, and a bowling alley. (WGS.)

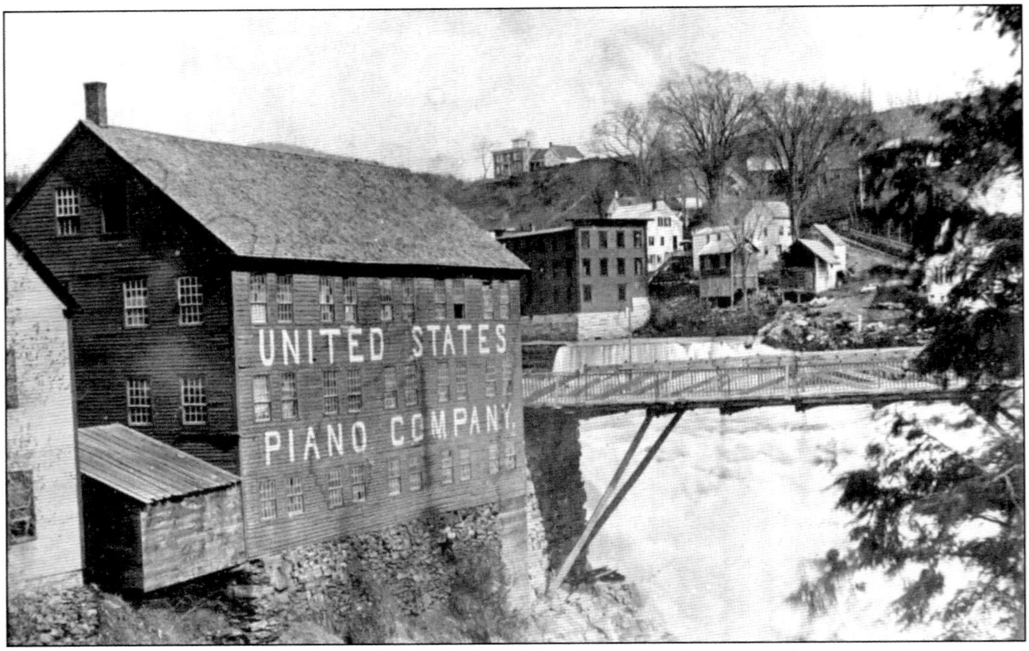


In many photographs, this building has no identifying signs. In this image, however, it has United States Piano Company painted on the side. A.J. Fullam's biography reads that he started a piano business in 1868 and moved it to New York City. There is no record of pianos being made here, so Fullam may have been using the building as a billboard. The building was owned by Slack & Ellison in 1871. (STL.)

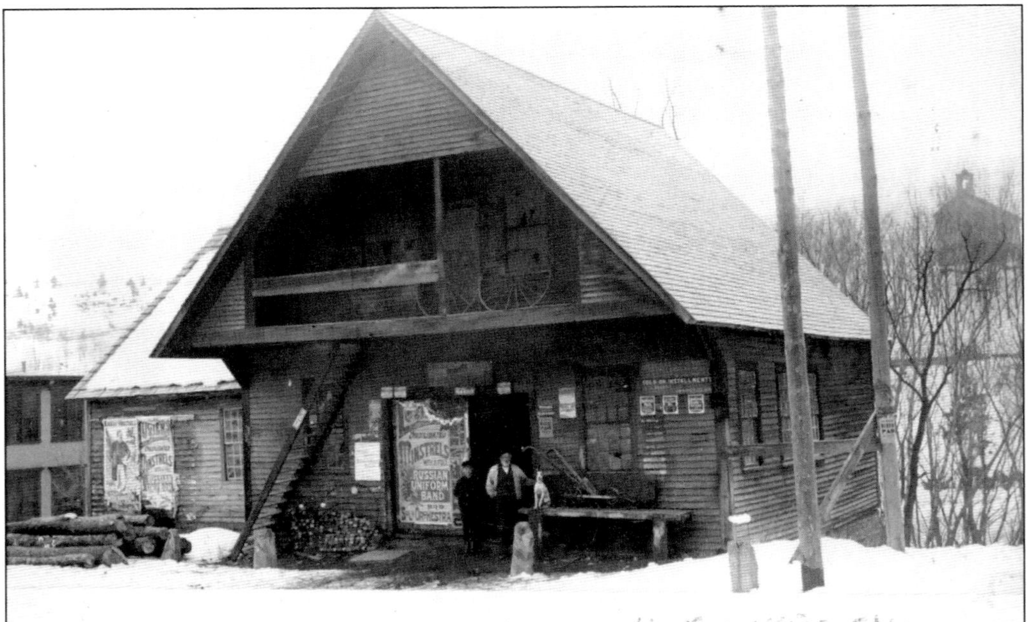


In 1828, Capt. George Kimball built this blacksmith shop near where Lovejoy Tool Company is today. In addition to the blacksmith shop, George's brother Brooks built horse rakes here. Kimball was succeeded by his son George, who would continue the business until 1890. The 1888 J&L building can be seen on the far left. (WGS.)

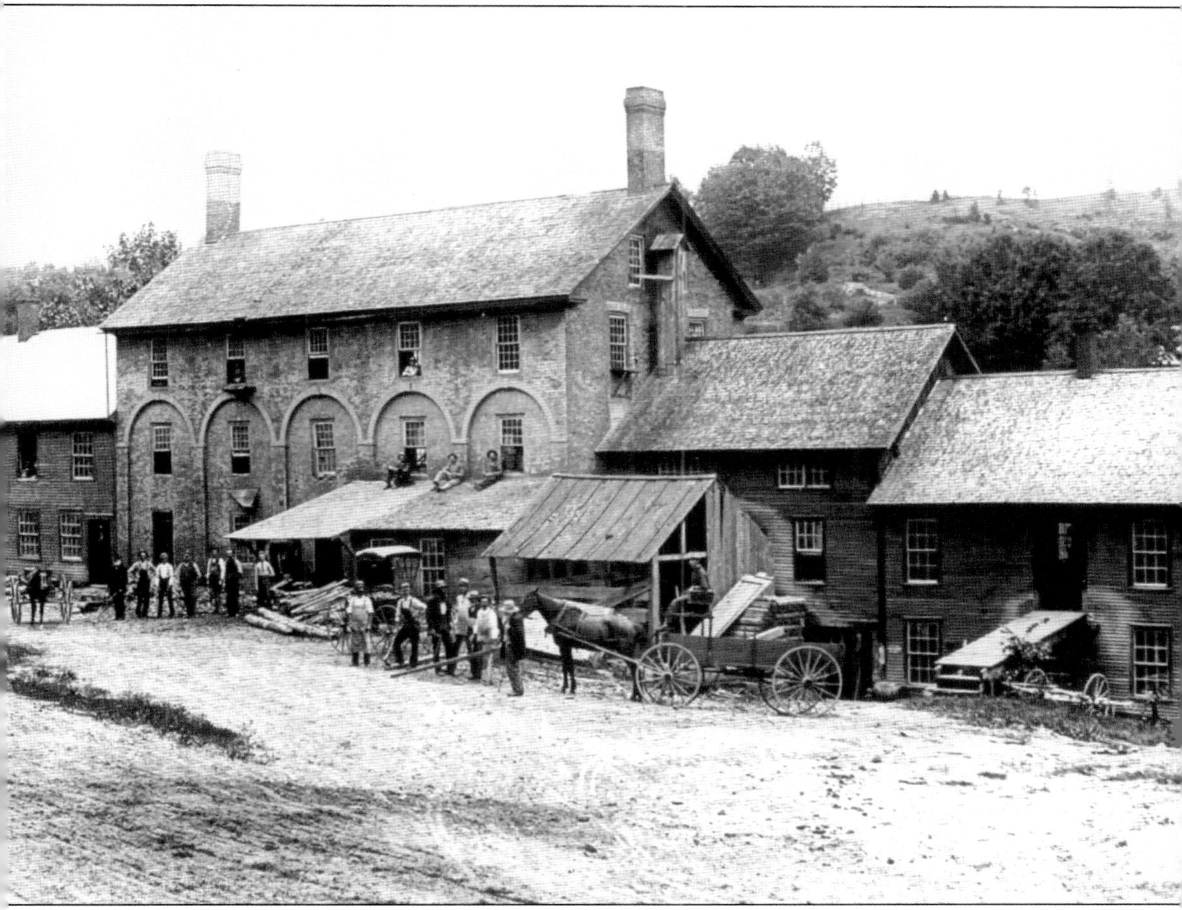

Col. Jonathan Williams came to Springfield in 1796. Around 1800, he built and ran a hotel on Main Street for a time. He had a tavern in his house on Clinton Street and a hat shop on Main Street. In 1812, he built this woolen mill (located where Lovejoy Tool is today), where he, and later his son, manufactured cloth. The building would house a variety of businesses over the years. The lower left was occupied by the Smith & Burr Hame Shop and the Mason & Davis Machine Shop. (Mason & Davis manufactured blanket hooks and eyes used by the army.) The large building in the center was occupied by Luke Taylor's mop shop, and prize churns were manufactured in the third and fourth buildings to the right. This site was purchased in 1888 for the original J&L building. (WGS.)

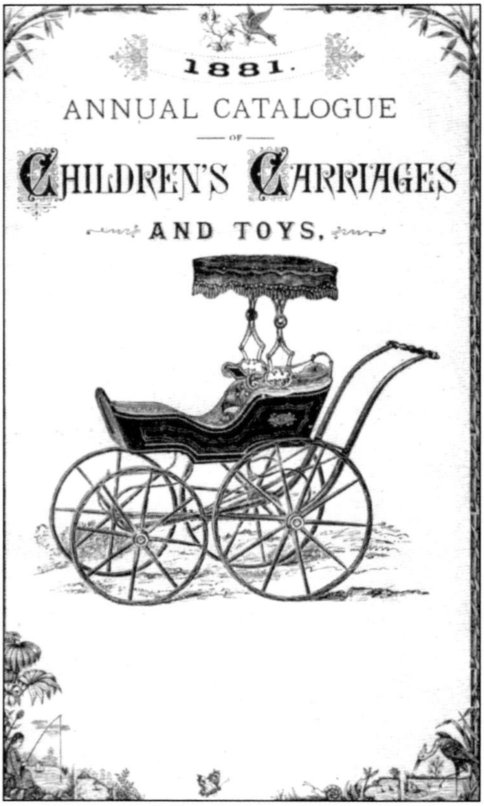


In the winter of 1856 and 1857, Joel A.H. Ellis developed a children's carriage, or cab. He made a sample and went to Boston and New York, taking orders for 1,500 carriages. He also invented the toy wagon and jointed wooden dolls. He was the first manufacturer of guitar and violin cases in America. This image is the cover of the 1888 catalog. The company was located in cab hollow, or Mineral Street. (MAC.)

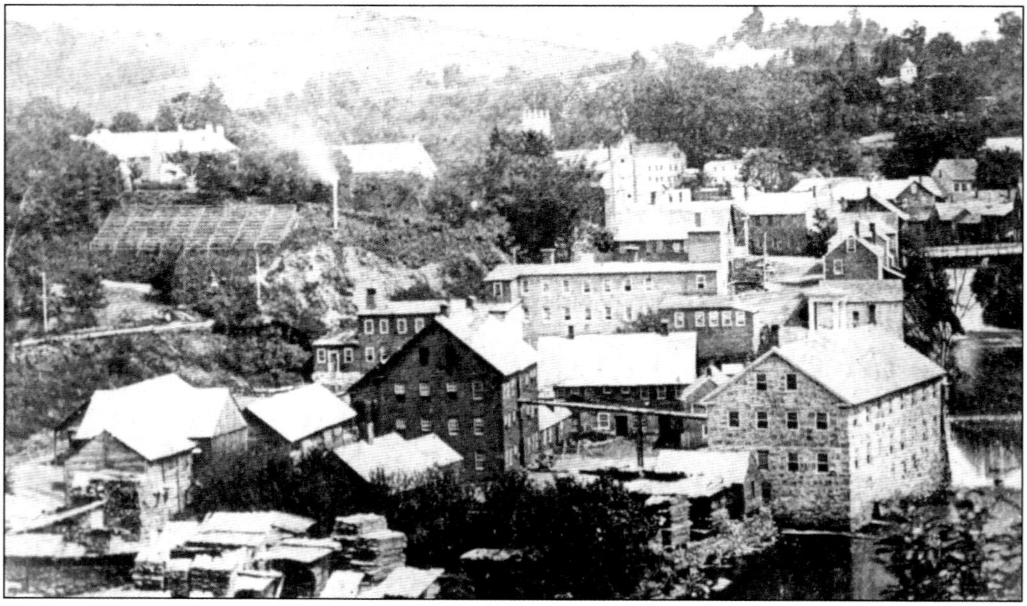

Herbert Ellis, in *Memories of Boyhood in Springfield, Vermont*, described the construction of the first dam below the Falls Bridge in 1863. Joel Ellis, his father, supervised the project and then erected a three-story stone building to house his carriage (cab) and toy business. Notice the second-story walkway connecting the stone building to the wooden one behind it. (WGS.)

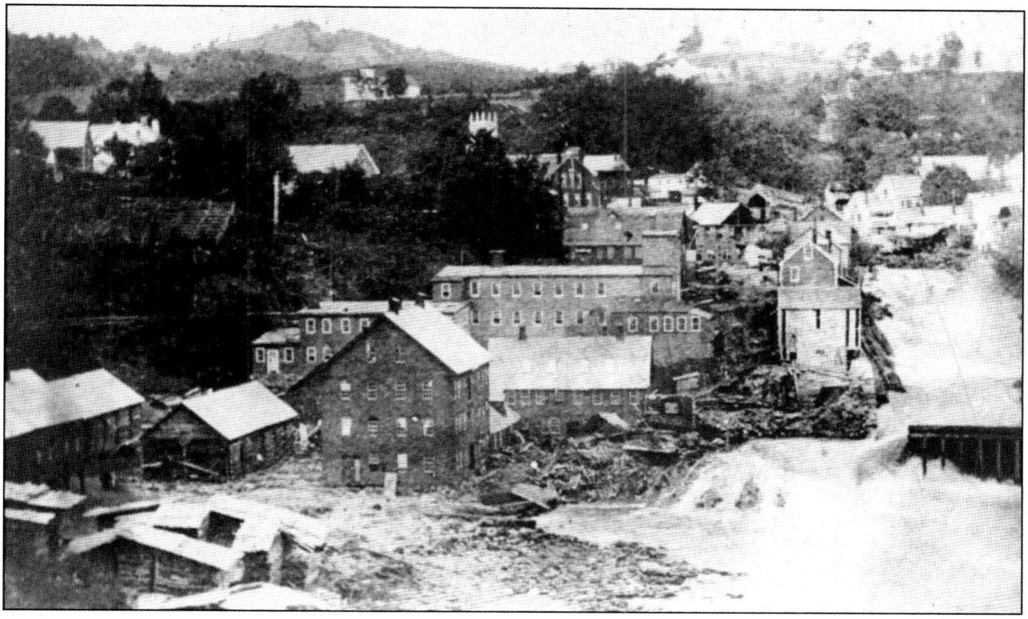

Unfortunately, only six years later, the flood of 1869 would take the beautiful stone building, the lumber and machinery, and many of the associated structures down the river. Herbert Ellis tells of his father going to the shop during the flood to cut holes in the floor in an effort to keep the water from washing the building away. The dam and business were rebuilt and opened the following spring. (WGS.)

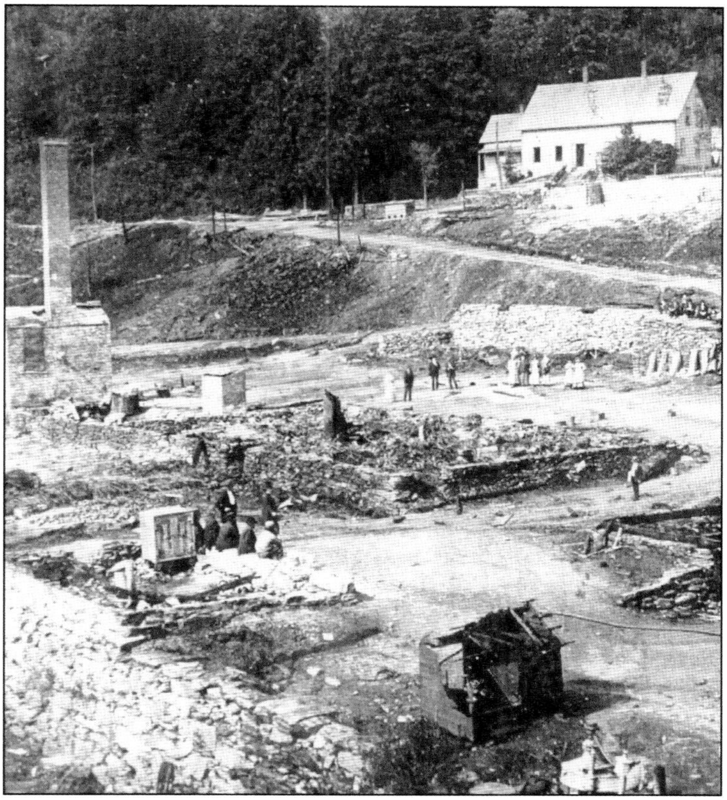

On June 25, 1878, at 7:10 p.m., the biggest fire in the history of the town broke out at the new cab shop. The wooden buildings, wood and shavings, and the sawdust inside combined with the hundreds of thousands of feet of lumber outside to result in what was described by Hubbard and Dartt as "a bonfire too hot to get near." By 9:00 p.m., it was a total loss. (WGS.)

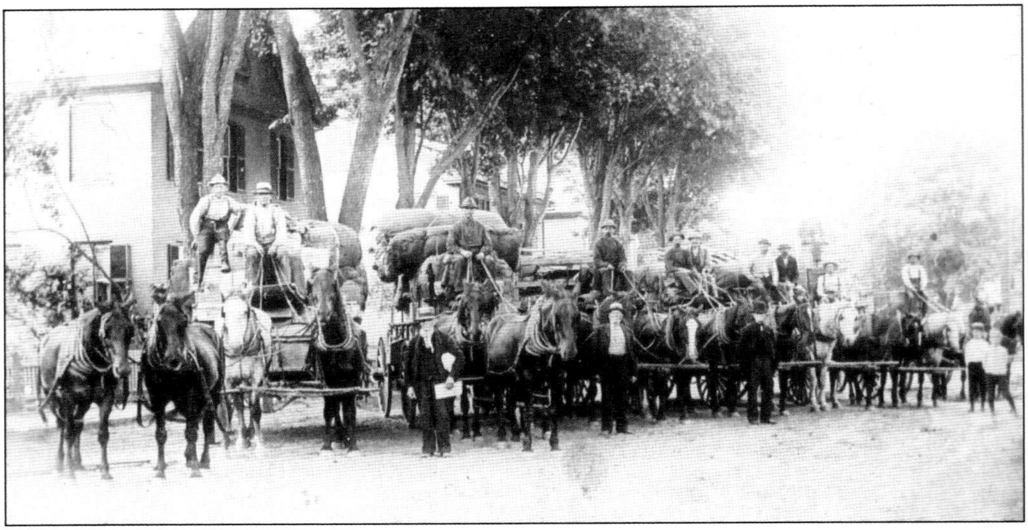

A load of shoddy is on its way to the mill in this image. Industry in the valley was growing, and this method of transportation was not very efficient. There were train stations in Bellows Falls; Gassets; and Charlestown, New Hampshire, but for many years there were only debates and failed plans to bring a train to Springfield. (WGS.)

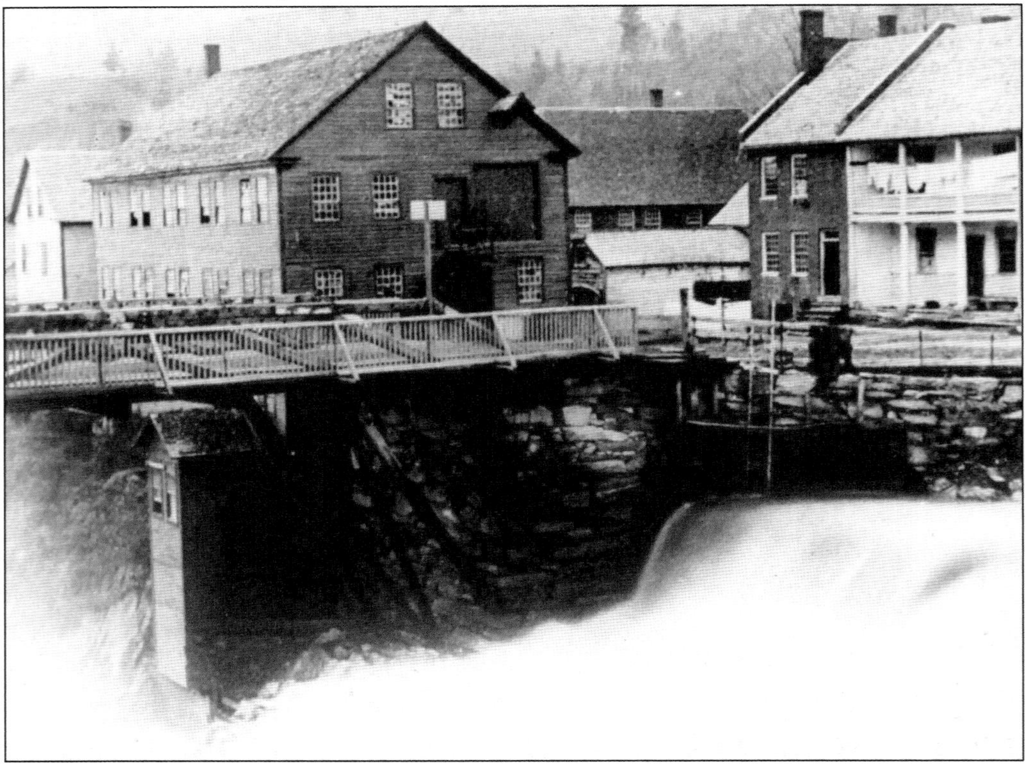

In the early 1800s, this building housed John Holmes's lead pipe factory and Horace E. Hawkins's furniture shop in the upper story. In 1871, W.H.H. Slack and F.G. Ellison, under the name of Slack & Ellison, started a shoddy business here. In 1916, this building and a line of wooden tenements along the river were torn down. Later, the current, 200-foot-long brick building was constructed. (WGS.)

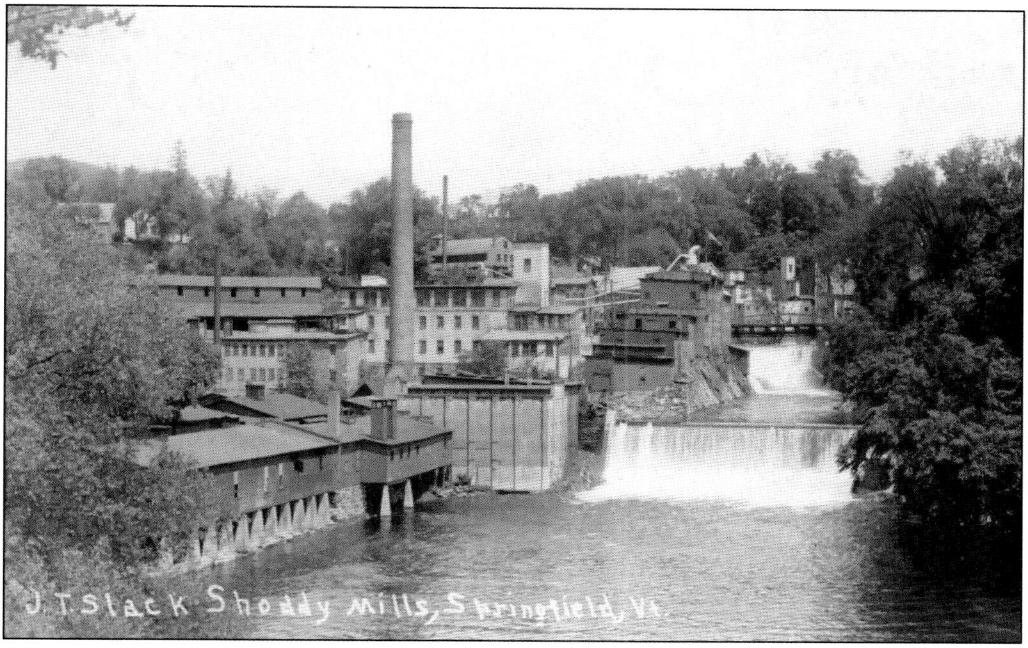


In 1911, W.H.H. Slack and John T. Slack, who had been running two separate shoddy operations, combined operations again. In 1918, they installed a new carbonizer plant that used muriatic acid to remove the cotton. The fumes discolored the hillside behind the plant. (HSP.)

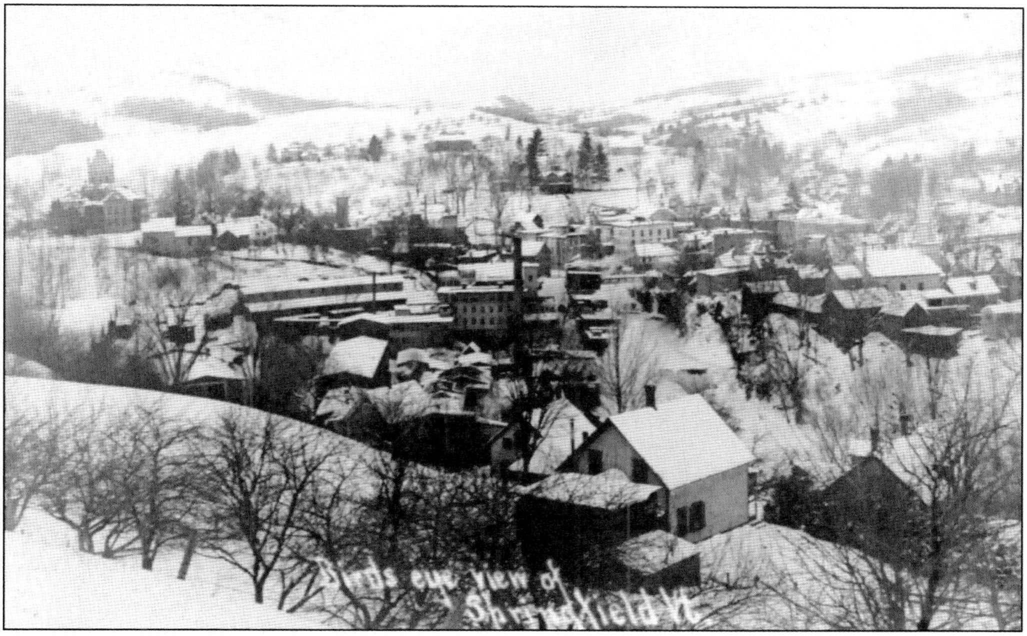


By 1922, the mill occupied most of the buildings seen on the left side of this photograph, including the old cotton mill, and employed about 325 people. The coming of synthetic fibers and a law requiring shoddy to be labeled "reused" were the beginning of the end. In and out of receivership in the 1930s, by 1951, the market was gone. In 1953, the mill declared bankruptcy and sold everything at auction. (STL.)

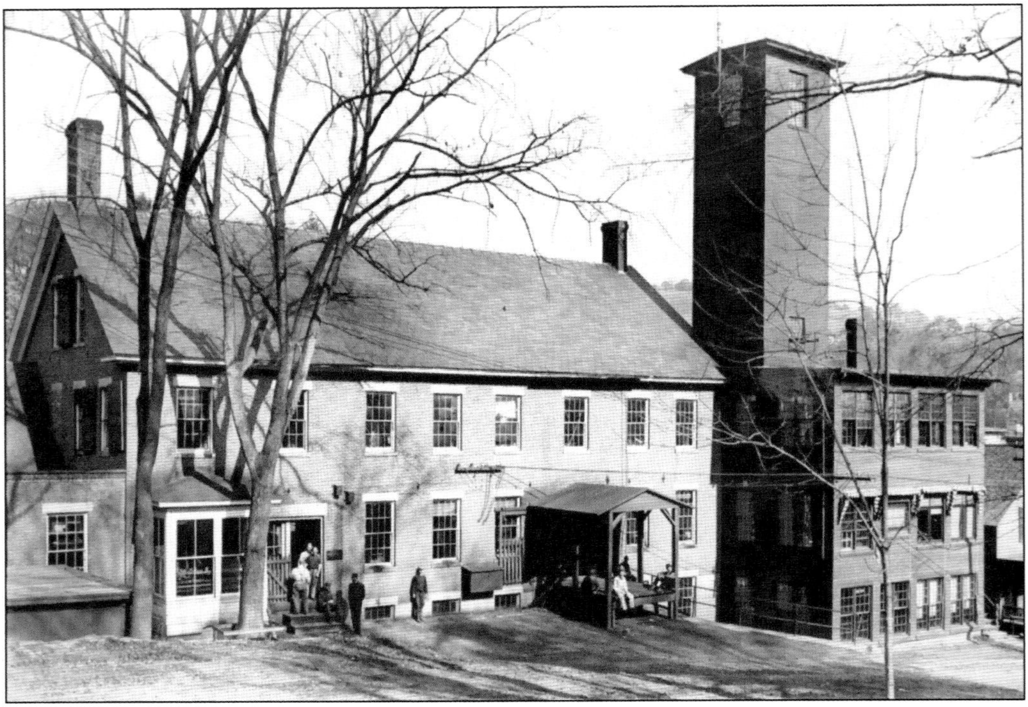

John Davidson and Frederick Parks began manufacturing shearing and other cloth-finishing machinery in 1829. The business was started in the brick building on the left. In 1846, Amasa Woolson joined the company. It became Parks & Woolson in 1850 when Davidson died. It was one of the oldest manufacturing businesses in town. (Lovejoy Tool Company.)

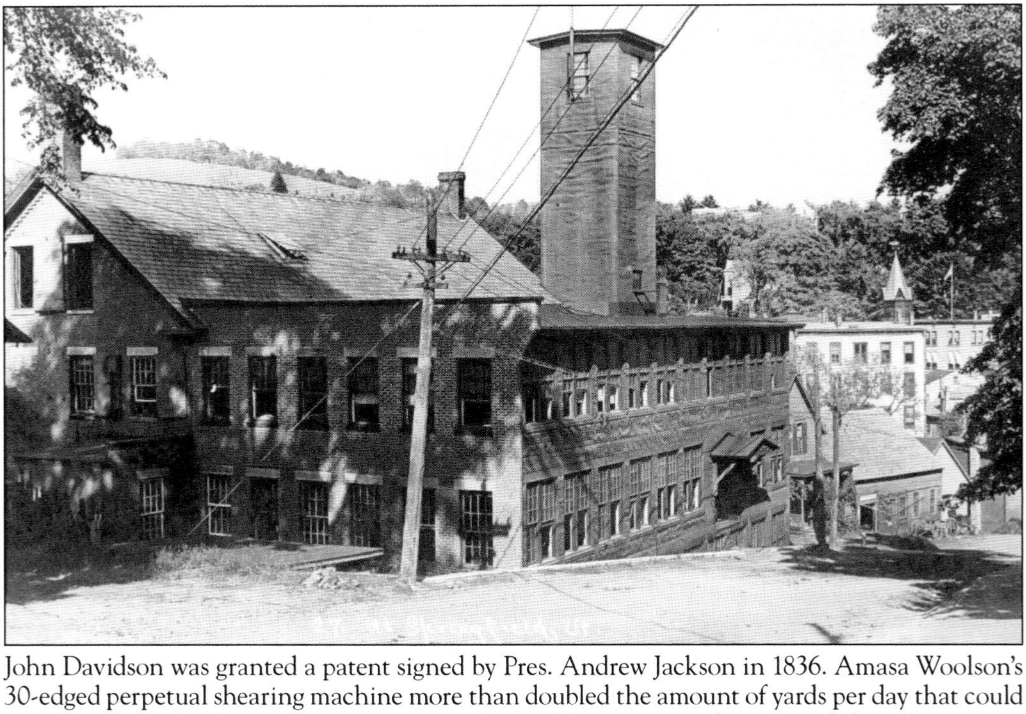

John Davidson was granted a patent signed by Pres. Andrew Jackson in 1836. Amasa Woolson's 30-edged perpetual shearing machine more than doubled the amount of yards per day that could be sheared (finished). As the business grew, the building continued to expand. (STL.)

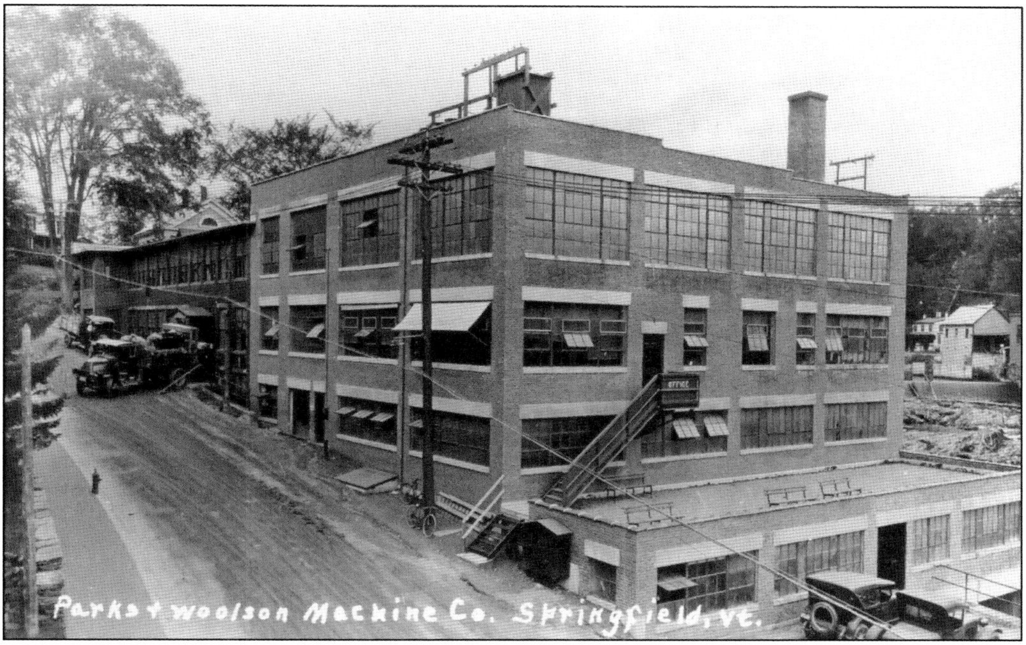


The building here appears much as it does today in this photograph from the 1920s. The original brick building has all but disappeared behind one addition after another. Many of the town's early leaders, such as Adna Brown, W.H.H. Slack, Frederick Ellison, and Farley Gilman, started their careers at Parks & Woolson. (STL.)

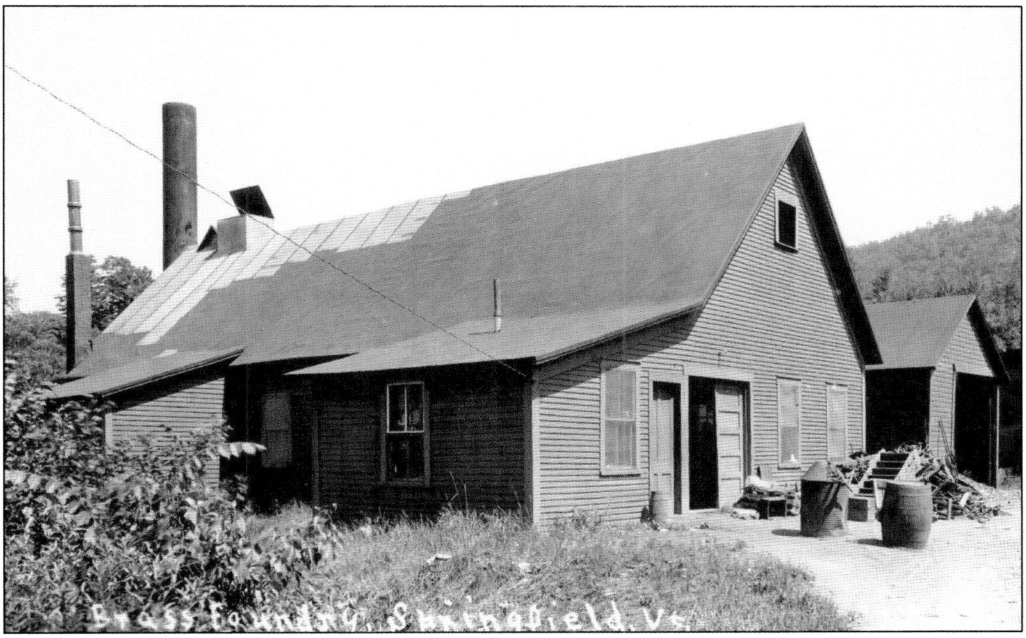


The Brass Foundry was located on the east side of the Black River by Seaver's Brook and was accessed by a bridge that crossed the river. The building also may have been accessible from Commonwealth Avenue. In early 1927, Frank Corliss tore down the old buildings and put up tourist cabins. Unfortunately, the November 1927 flood washed away both the bridge and the cabins. (STL.)

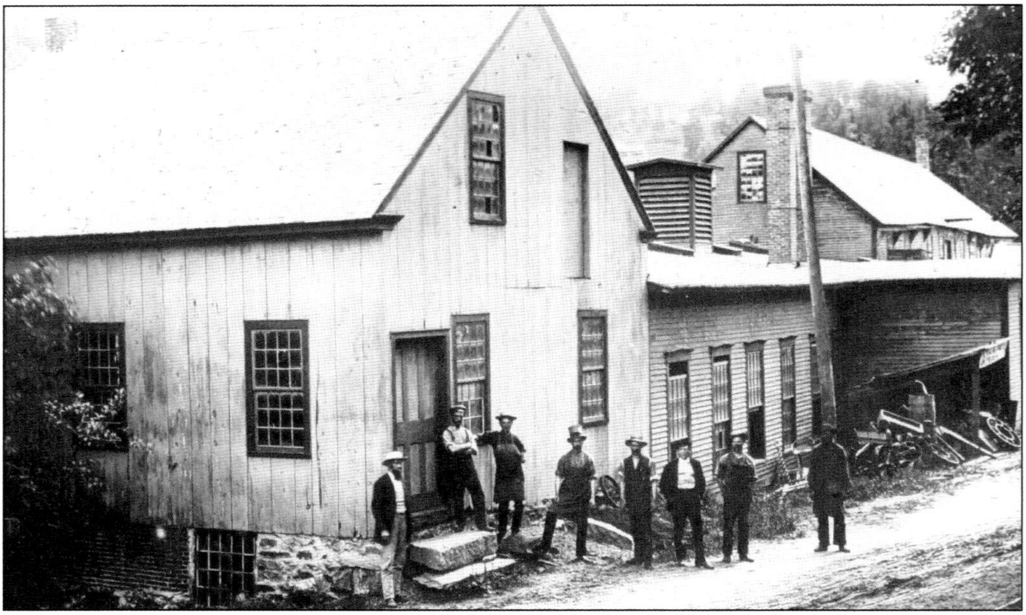

The Vermont Snath Company (a snath is the handle to a scythe) was organized in 1869 and located in a group of buildings on Clinton Street by Bridge Street (roughly where the recreation center is now). The firm also cast metal. When the snath business was moved to Bellows Falls in 1882, the building became a foundry. (WGS.)

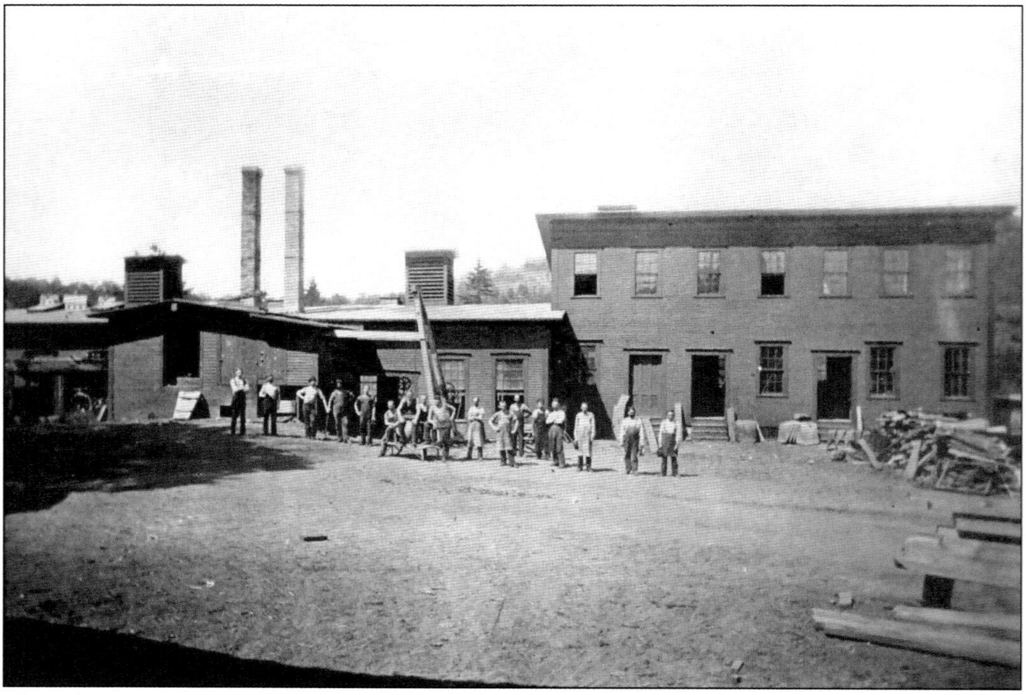

Though still called the Vermont Snath Company, by 1913, the company was generally in line with what the other businesses in town were doing. By 1917, because of increased business, it expanded across the river. In 1931, the Vermont Snath Company was sold to Vermont Foundries, a company run primarily by J&L and Fellows Gear Shaper. (WGS.)

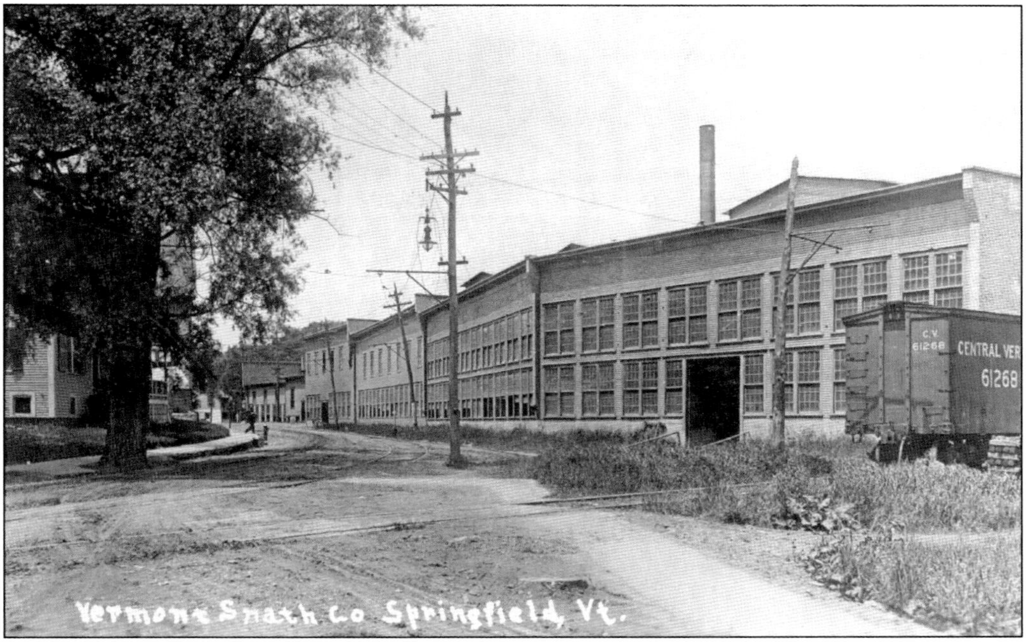


Though not technically a machine tool business, much of the Vermont Foundry's production was the direct result of machine tool orders. In 1947, it had a shut down due to lack of materials; however, by 1949, it had 362 employees. In six years, that number would shrink to only 150. Because of its ties to the machine industry, the closing of three of the four major companies spelled the end of the foundry. (HSP.)

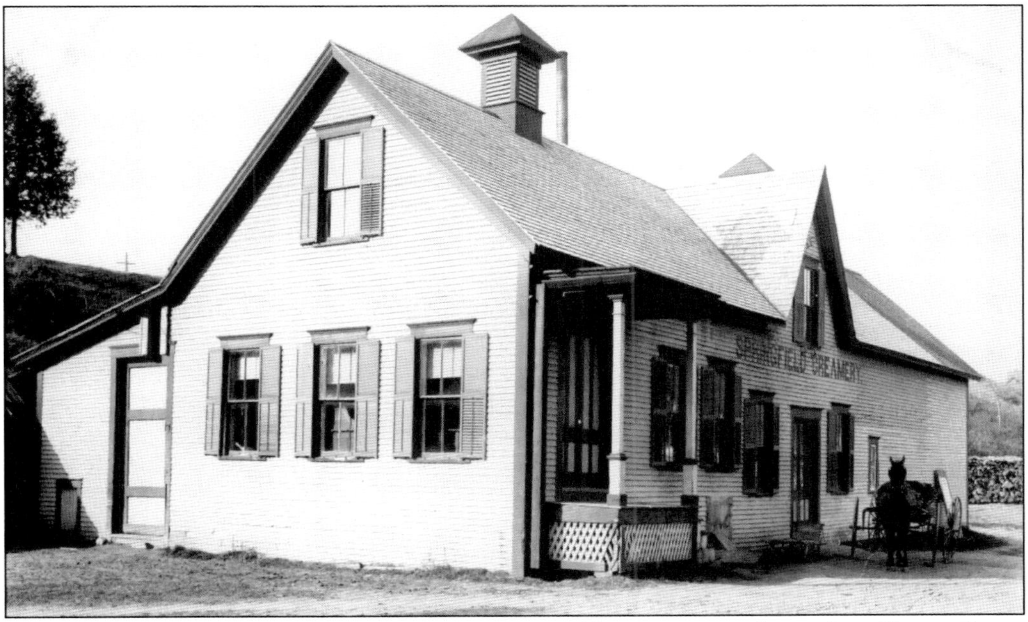


Although this photograph is hard to date, the Springfield Creamery was located on Clinton Street just south of Furnace Street, possibly about where the gas company was later located. There was also a creamery in North Springfield. In the hills there were many farms milking large dairy herds. At one time, the town had three milk producers—Hastings, Estey, and Idlenot—but they are all gone now. (STL.)

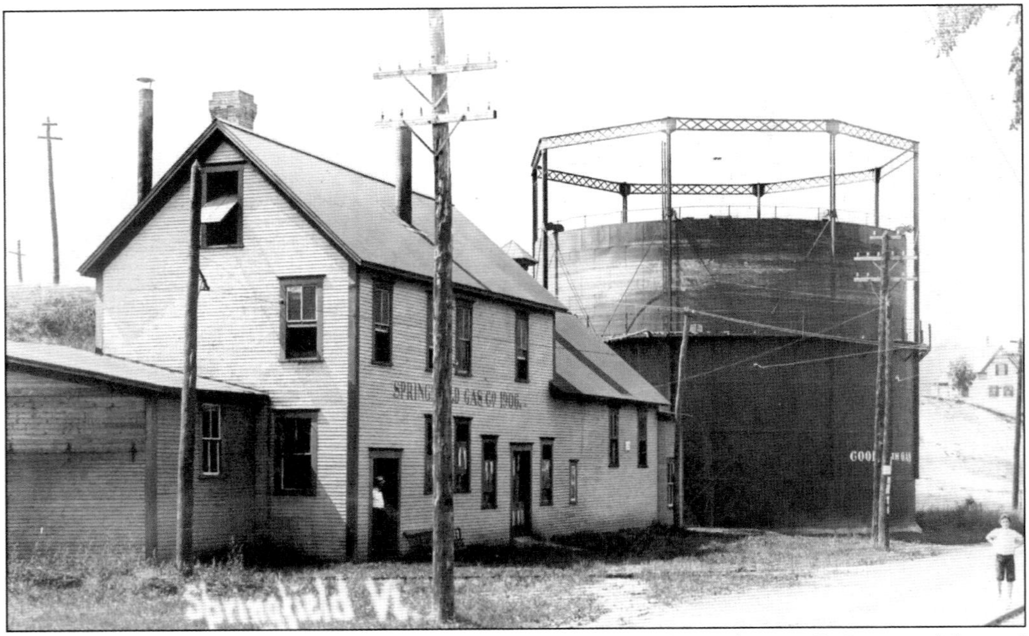

Located on Clinton Street, the Springfield Gas Company was formed in 1905. Gas was delivered through a quarter (25¢) meter; when the flame was low, it was time to put in more quarters. Initially used for cooking and lights, gas was too expensive for heat. When everyone was cooking at the same time, pressure could be a problem. Gas was locally manufactured until it was replaced by imported propane in 1950. (STL.)

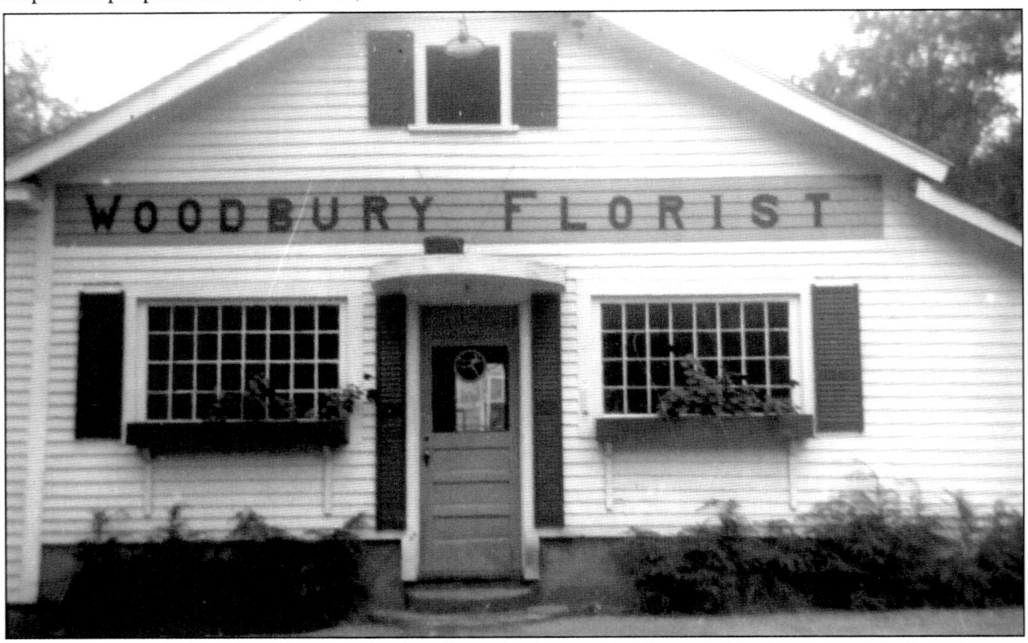

In June 1946, with no experience in the floral business and four children in tow, Karl and Berta Woodbury bought May's Flower Shop (and greenhouses) on Perley Gordon Road. At that time, greenhouses were used solely for growing flowers—like coaxing hundreds of Easter lilies to open at just the right time. After two Main Street locations, the store on River Street was opened in 1975. (Alan Woodbury.)

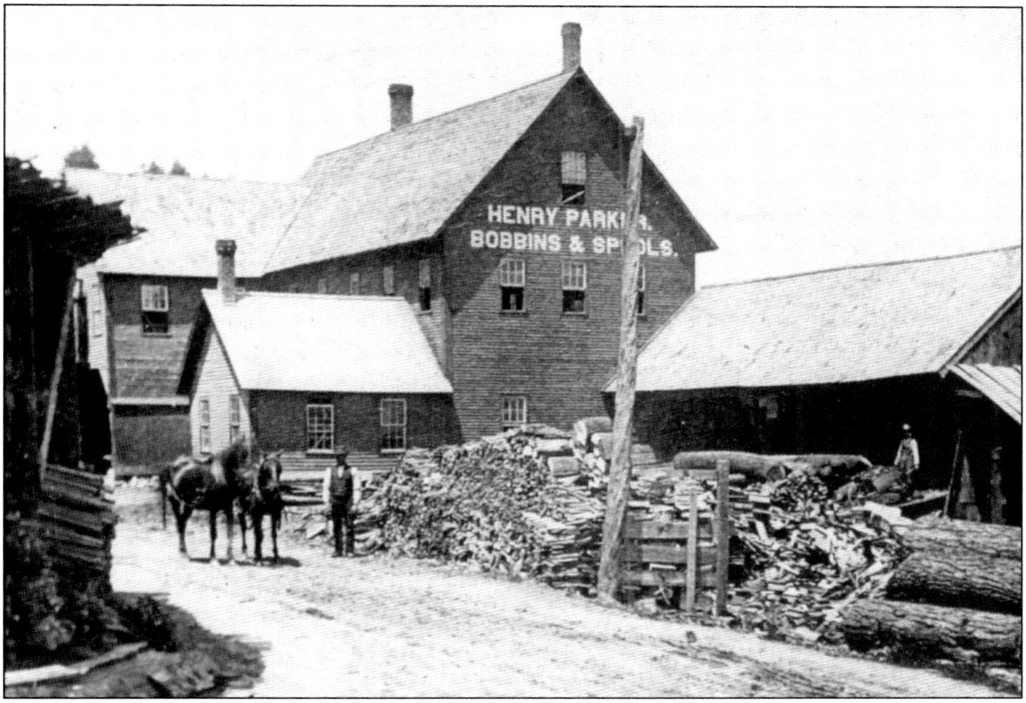

Henry Parker and Alphonso Washburn established the Henry Parker Company in 1864. The company made bobbins and spools. Located at the foot of Randall Hill in the Goulds Mills area, it had been the location of a sawmill. The site would later be used for a hydroelectric dam that primarily powered the trolley. (WGS.)

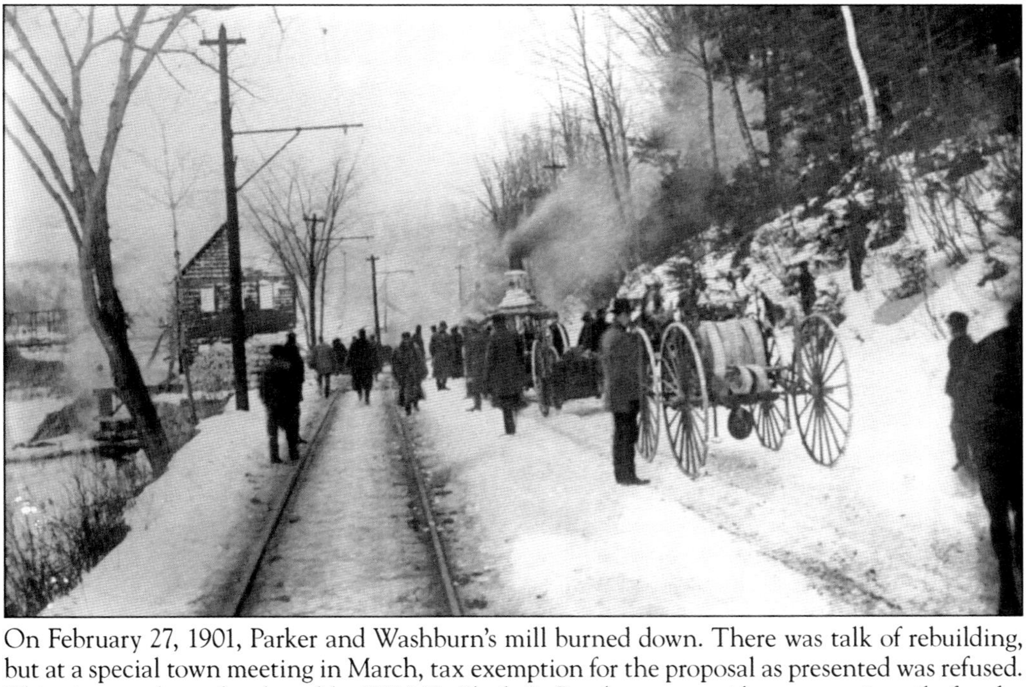

On February 27, 1901, Parker and Washburn's mill burned down. There was talk of rebuilding, but at a special town meeting in March, tax exemption for the proposal as presented was refused. This site was later developed by W.H.H. Slack & Brother to provide power primarily for the electric railway. (WGS.)

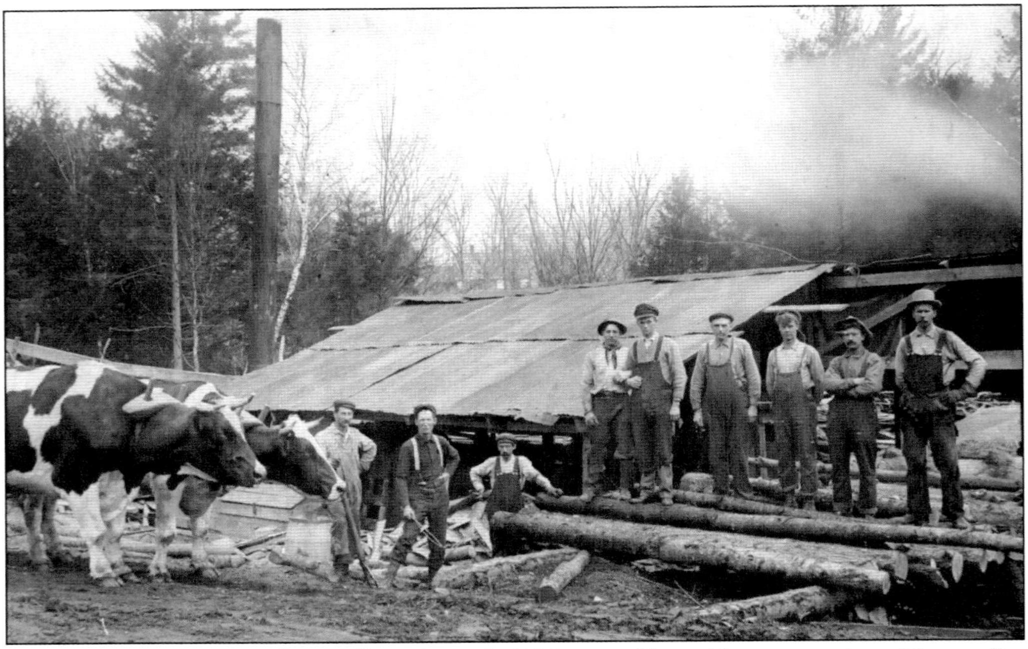

There were to two types of sawmills in the early 1900s: portable and large operations. The smaller local, or portable, sawmills were operated by farmers who earned extra income timbering their property or land nearby and sawing the logs into lumber to sell locally in the winter. In this photograph taken around 1914, men pose in front of Solon Putnam's sawmill on Putnam Road. Putnam is standing on the logs and is the second man from the left. (HSP.)

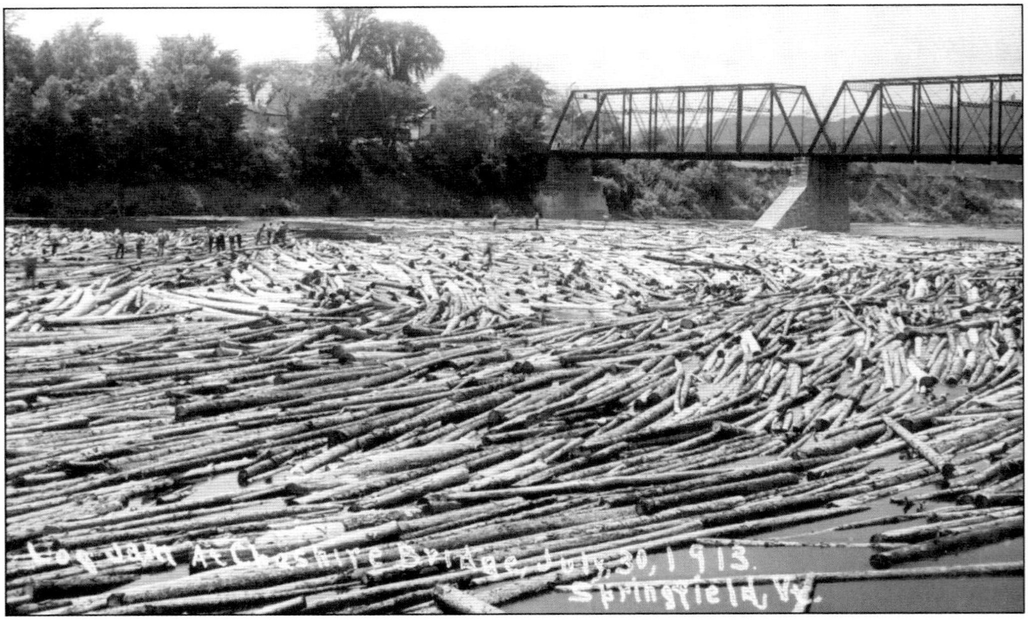


There were also large logging operations that were primarily located up north. This 1913 log drive on the Connecticut River is likely heading to Massachusetts or Connecticut. It was far easier to float the logs down the river to a sawmill than to move finished lumber by oxen and railroad. This was not a job for the faint of heart, however; many lives were lost breaking up logjams on the river. (HSP.)

## *Five*

# Machine Tool Industry

This chapter is about the Big Four: Jones & Lamson Machine Company, Fellows Gear Shaper Company, Bryant Chucking and Grinding Company, and Lovejoy Tool Company.

The textile business was crucial to the coming change to machine tools. The creativity and ingenuity at the shoddy mill and Parks & Woolson made these companies leaders in their fields. These crafts would now be used to build a highly skilled machine tool town.

In 1888, J&L's heavy machinery, which originally included seven product lines, was brought by oxen from Windsor over 17 miles of dirt roads and Skitchewaug Mountain. A new building would be erected on the old Williams Mill site by the Clinton Street Bridge.

The former Windsor plant manager only came for a short time, so a new man had to be hired. Fortunately, it turned out to be James Hartness from Torrington, Connecticut. J&L would be the start of the Big Four.

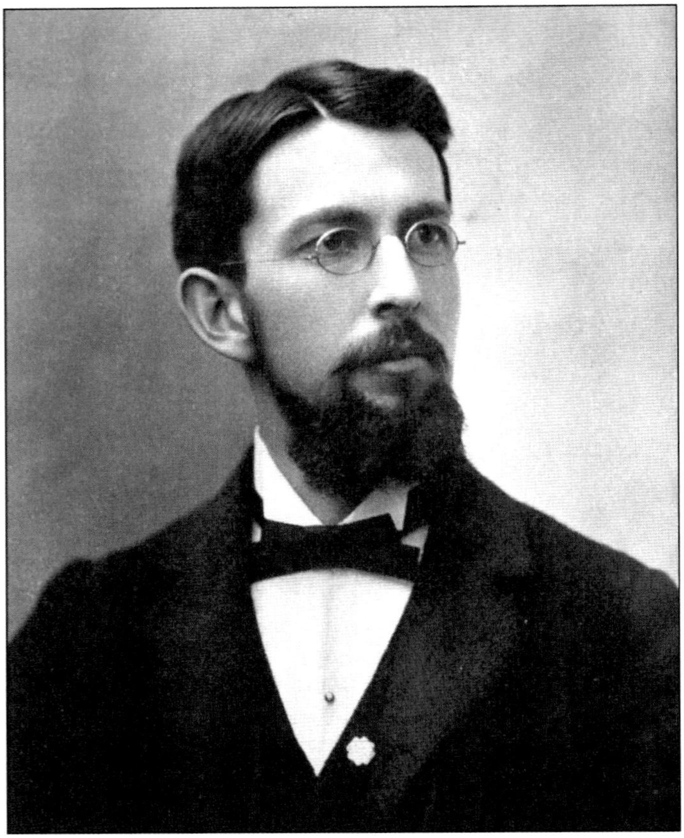

James Hartness was 27 years old when he came to Springfield to run J&L. Not only a creative genius and a great judge of people and ideas, he was also a visionary. When other people saw aviation as a rich man's sport, he saw the future of transportation. He would set the standard for the town for the next 45 years. (Hubbard and Dartt.)

Right away, Hartness abandoned all the products from Windsor except the high turret lathe, which he redesigned and patented in 1891. Renamed the Hartness Turret Lathe, it became the only J&L product. This is the first J&L building, photographed before the structures on either side were added. The short side on the right is where the front door is today. (STL.)

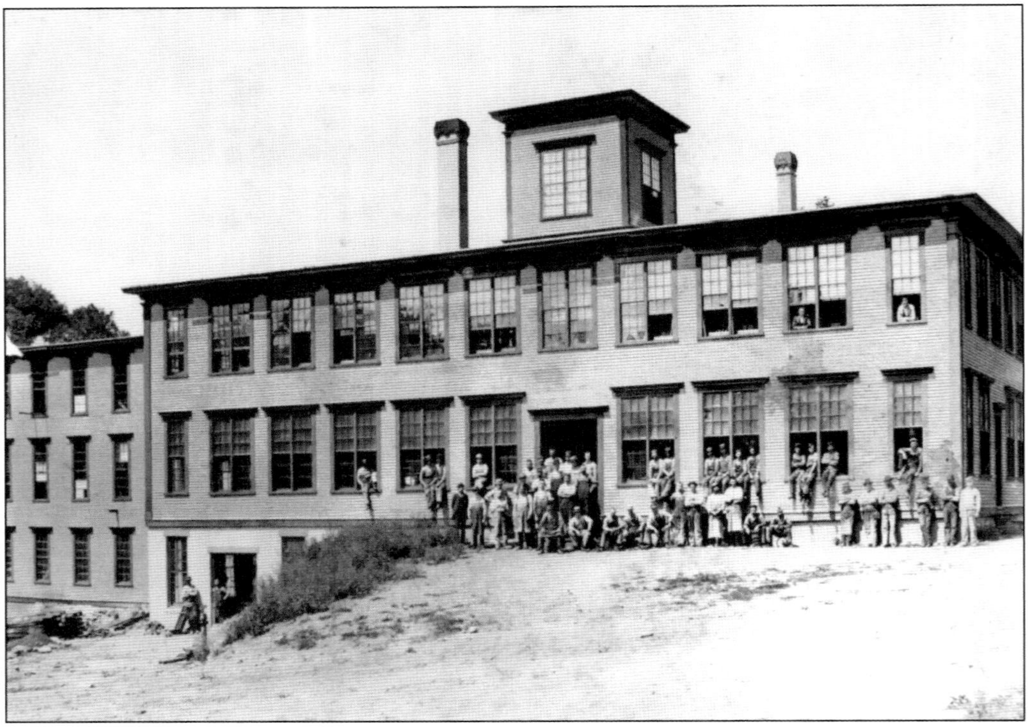

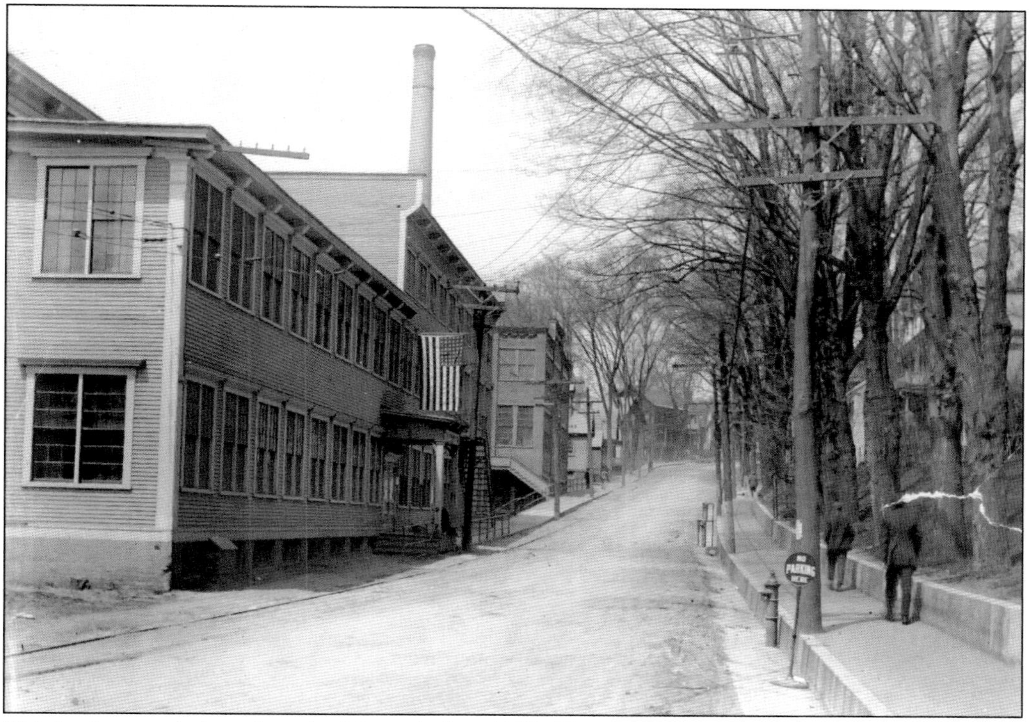

J&L had expanded to three buildings before 1907. These structures were commonly called the upper shop. However, three buildings were not enough. In 1907, J&L started construction on a new location on Clinton Street—the lower shop. Notice the covered entry over the door and how close the trolley tracks were to the building. (STL.)

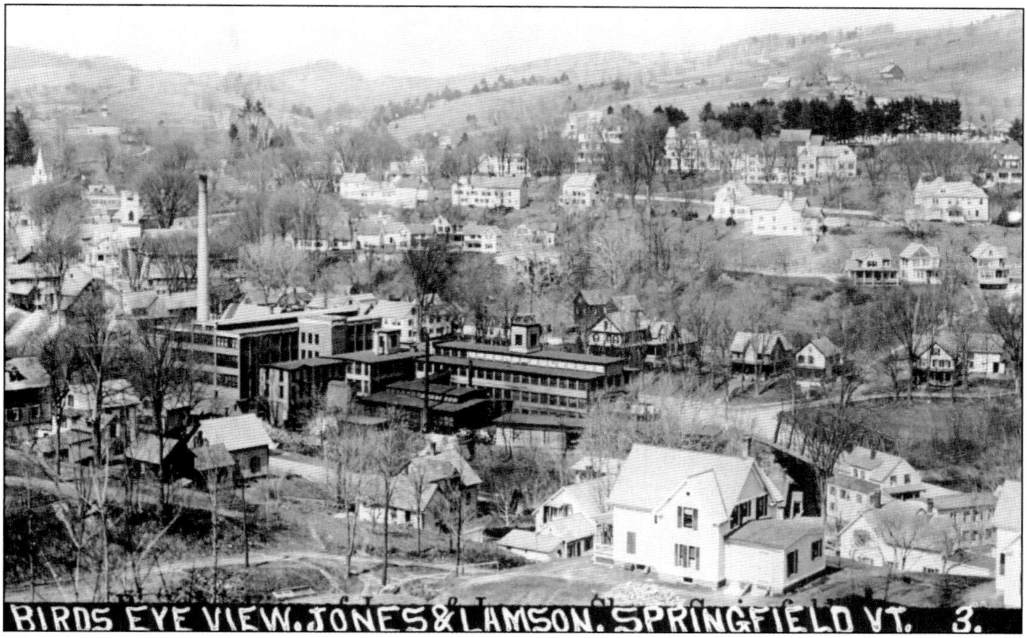


This photograph was taken some time after 1900 when the Episcopal Church was built and before 1911, the date on the postcard. One can see how J&L grew in the 20 years between 1889, when the original building was constructed, and when this photograph was taken. (HSP.)

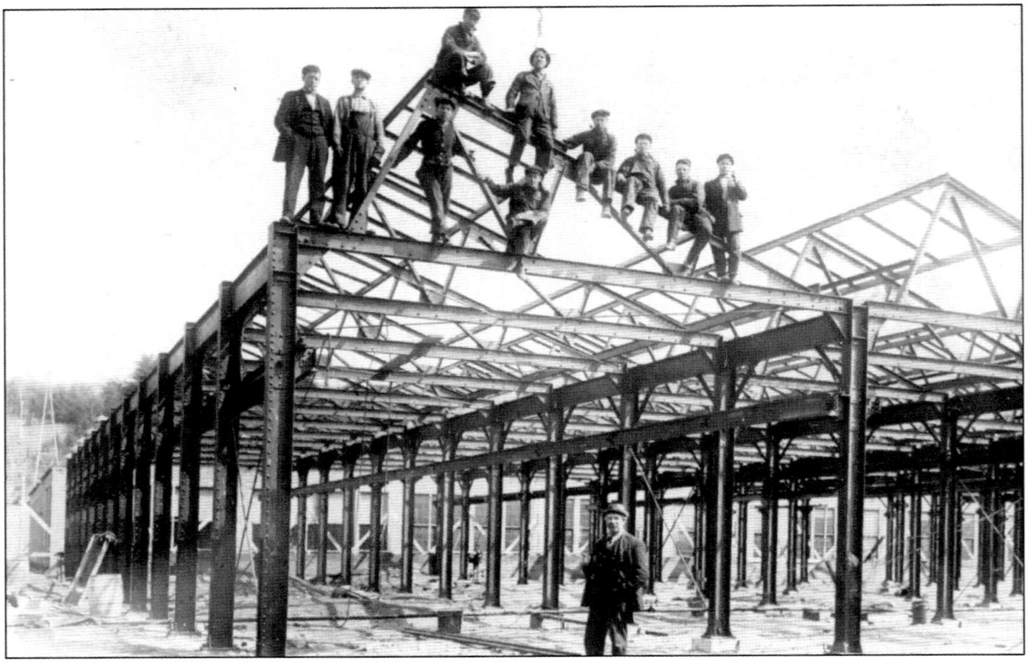

The building continued after the operation moved to Clinton Street. This was the second edifice, located between the original structure nearer the river and Clinton Street. The first building can be seen behind the workers. (STL.)

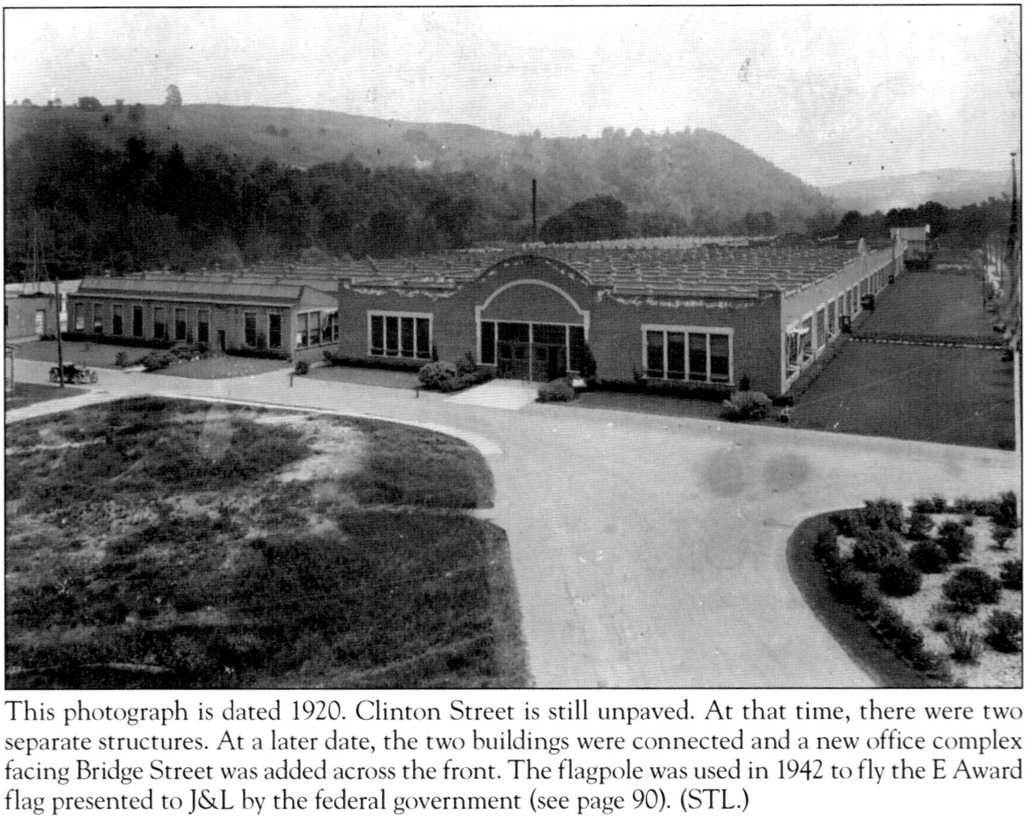

This photograph is dated 1920. Clinton Street is still unpaved. At that time, there were two separate structures. At a later date, the two buildings were connected and a new office complex facing Bridge Street was added across the front. The flagpole was used in 1942 to fly the E Award flag presented to J&L by the federal government (see page 90). (STL.)

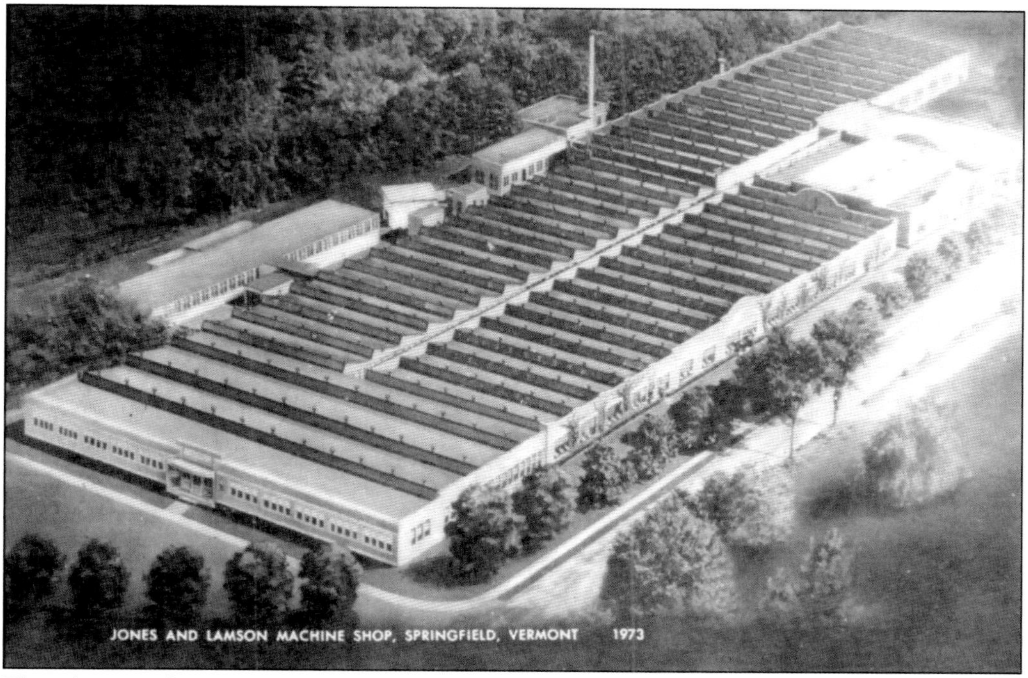


This photograph was most likely taken in the late 1940s. However, there was more construction after this. Additional factory space would be added to this building, and then Plant 2 would be erected. (HSP.)

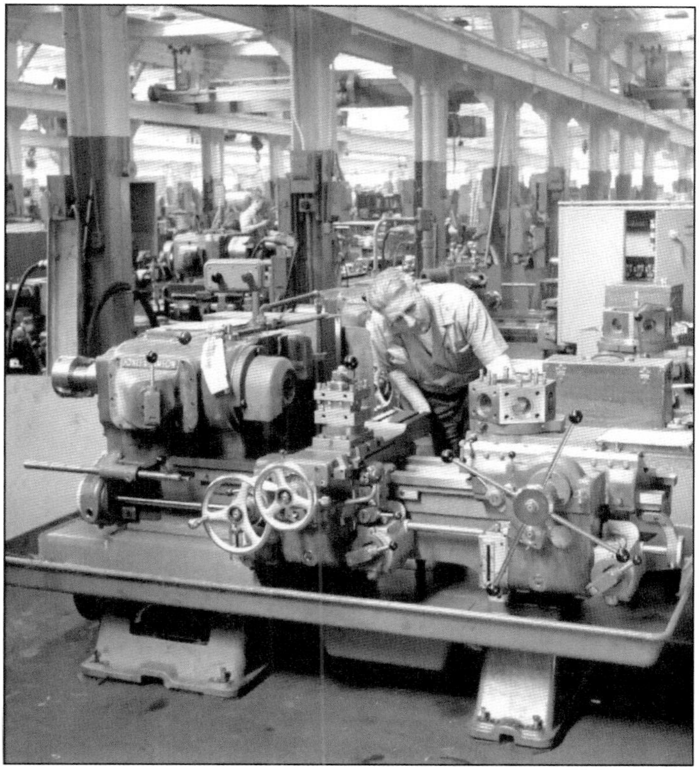

In this photograph, probably taken in the mid-1950s, Eugene Thibault is inspecting a flat turret lathe. Thibault, originally from Fitchburg, Massachusetts, was drawn to Springfield, like so many others in the late 1930s, by good job opportunities in the shops. (Author's collection.)

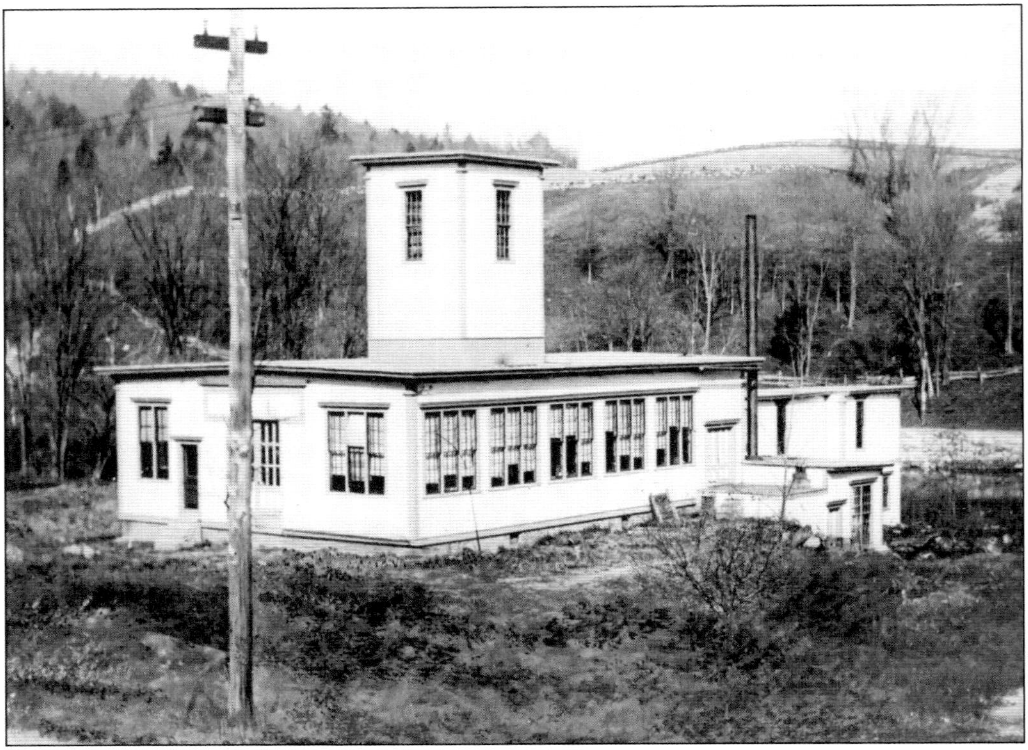

In 1897, Edwin Fellows—with the backing of James Hartness, Adna Brown, W.D. Woolson, and John Slack—started his new gear shaper business in this 45-by-60-foot building located on Pearl Street by the Black River. This was the first spin-off business from J&L. (WGS.)

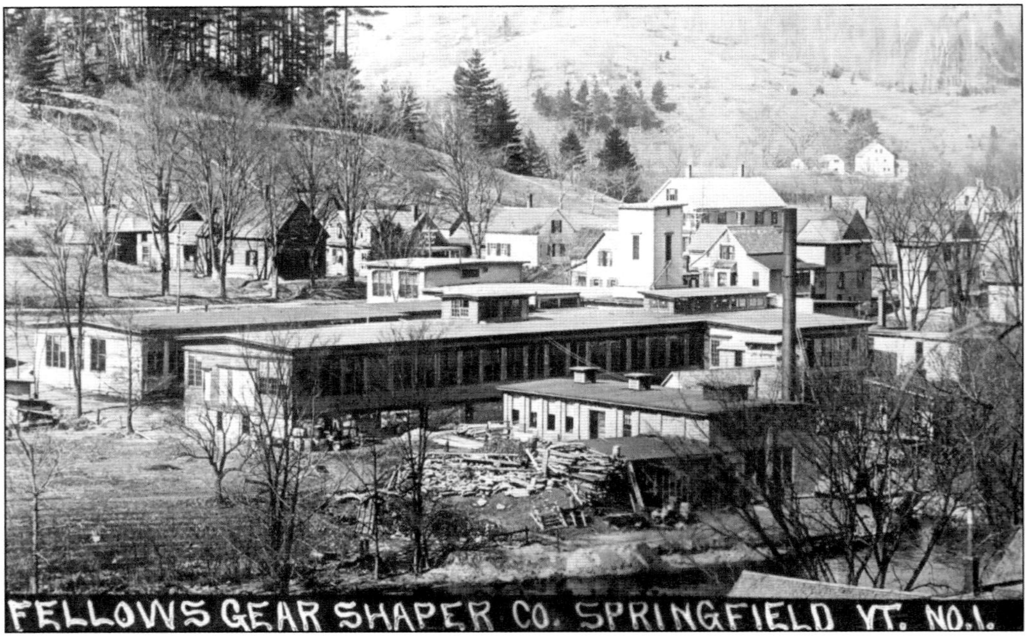


The gear shaper revolutionized the gear making industry. These machines made gears far more accurately than one could by hand, and it also made them faster. This was important for the burgeoning car industry and the airplane manufacturers that would follow. (STL.)

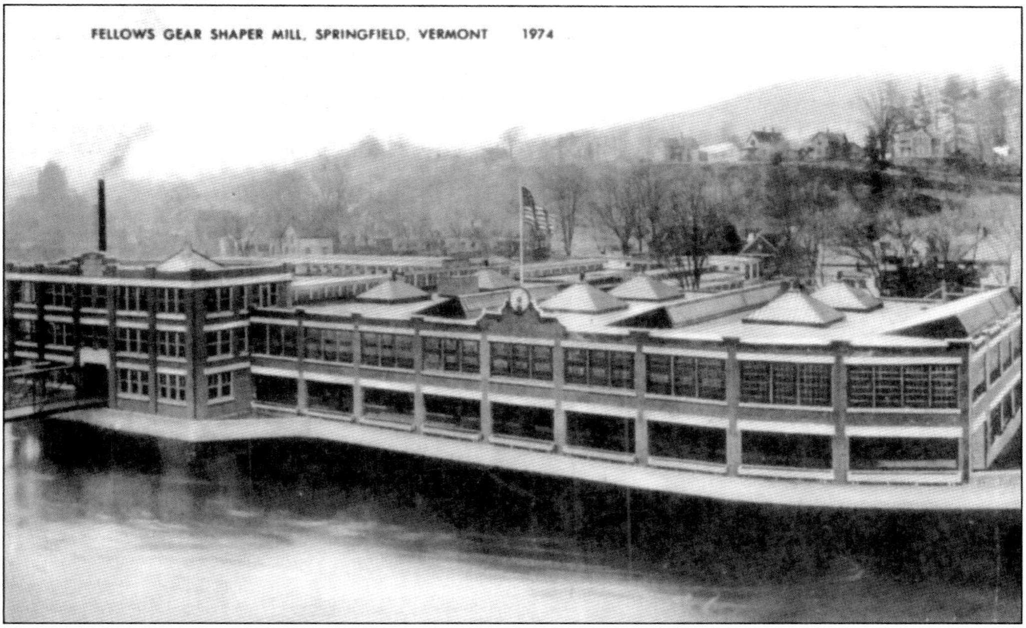


After World War II, along with gear shaping, Fellows would move into plastic injection molding machines. In this photograph from the 1920s, the building takes up a major part of the riverbank between the original upper dam and the beginning of the falls. As with J&L, the plant would expand, adding extra space and additional buildings. (HSP.)

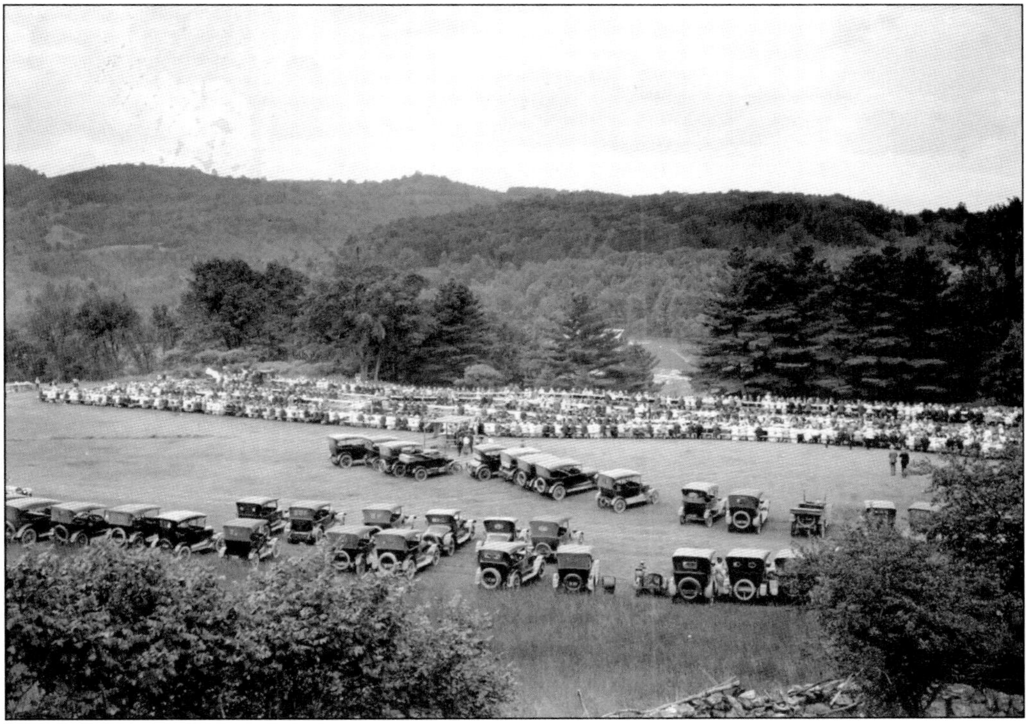

In 1919, E.R. Fellows rewarded his employees for producing 100 gear shapers and 1,000 cutters in one month. He threw a barbecue for them. The party was held on the site where Westview would be built during World War II. (STL.)

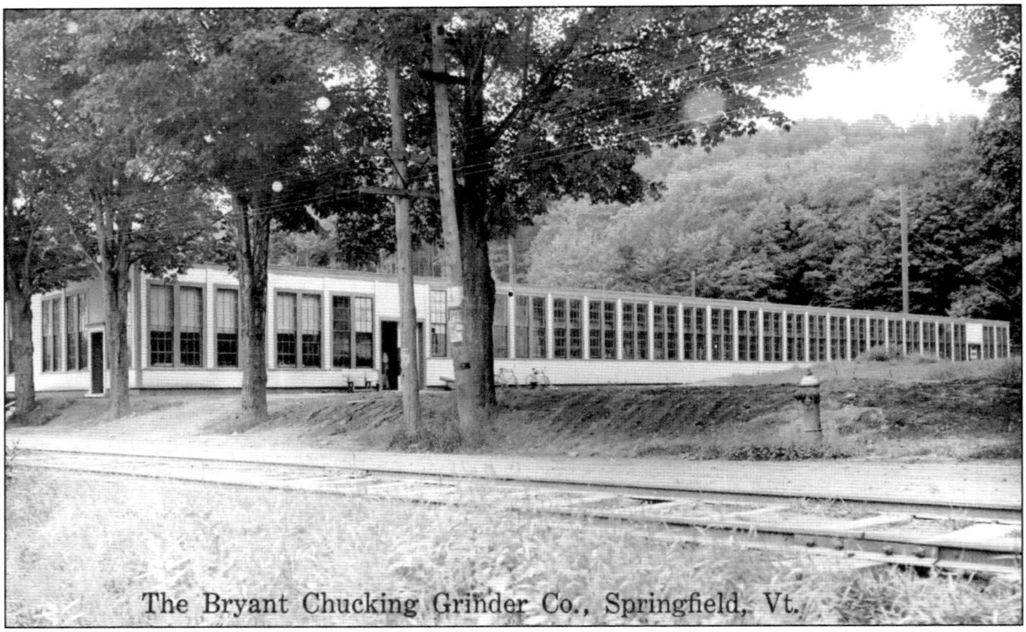


William Bryant, who had been in charge of the lower shop–building project, invented a new chucking grinding machine. As with Fellows in 1910, James Hartness and W.D. Woolson set up William Bryant and his new machine in this 40-by-230-foot building located across the street from J&L. Hartness still firmly believed that a company should only make one product. (HSP.)

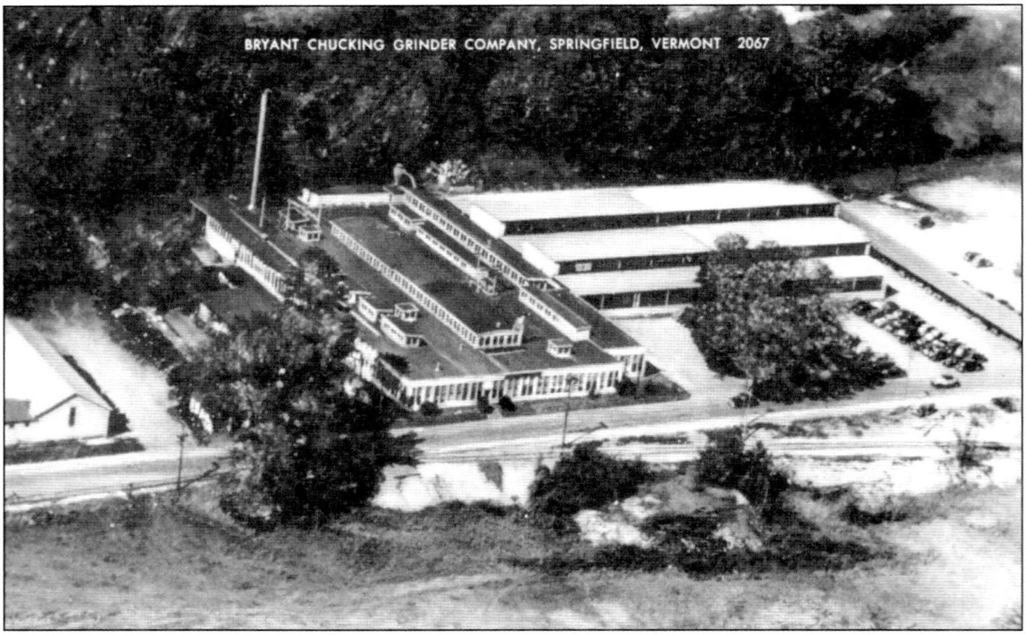


It is difficult to tell where the original Bryant building was in this photograph probably taken in the late 1920s. As with the two other plants that were already in operation, expansion was a continuing event. Substantial additions were made on both ends of the building. (HSP.)

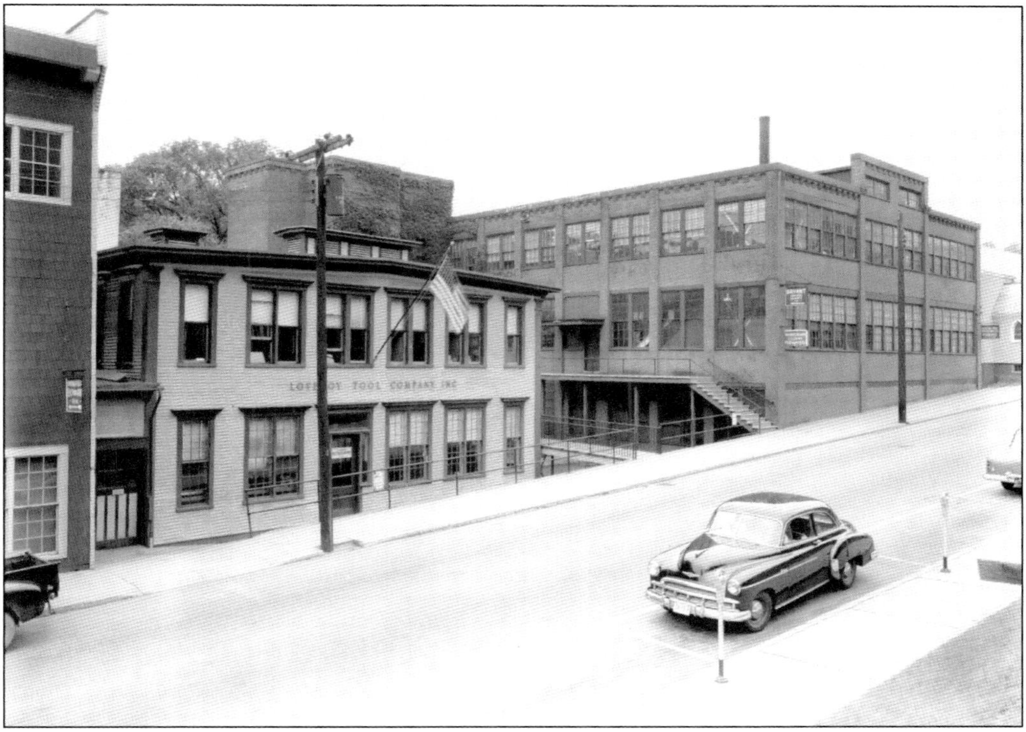

James Hartness had once again found a chief engineer who was also an inventor. Fred Lovejoy invented an inserted blade lathe tool that allowed the blade to be replaced instead of the whole tool. In 1917, as was done twice before, Hartness sent Lovejoy to the upper shop to set up and run his new business. (Lovejoy Tool Company.)

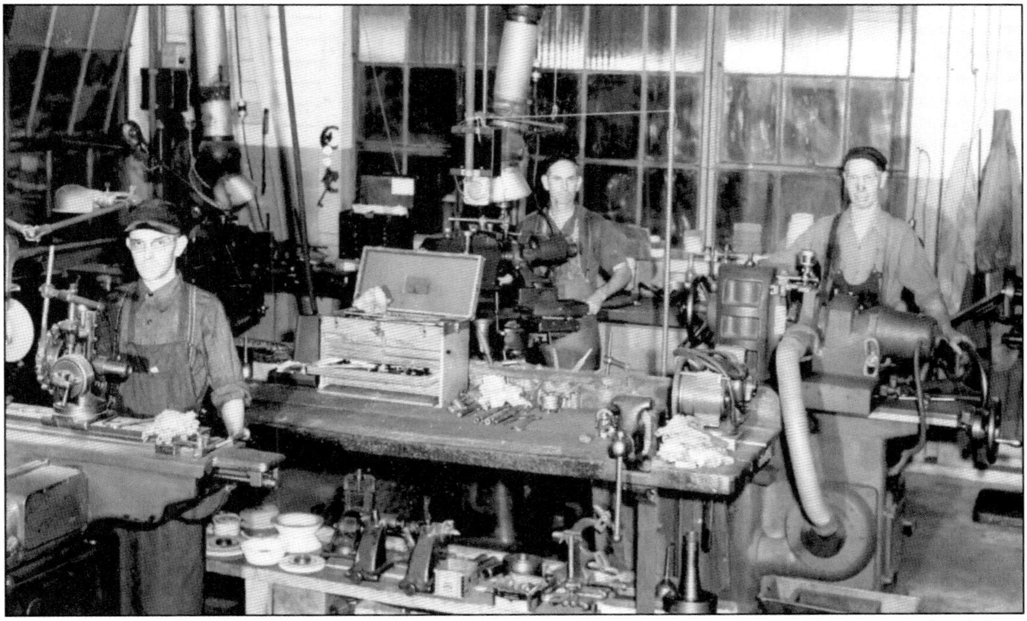

The Lovejoy cutter-grinder shop was photographed in the 1940s. The operator on the left is Harry Gardner, thought to be one of the original employees. The other two men are unidentified. (Lovejoy Tool Company.)

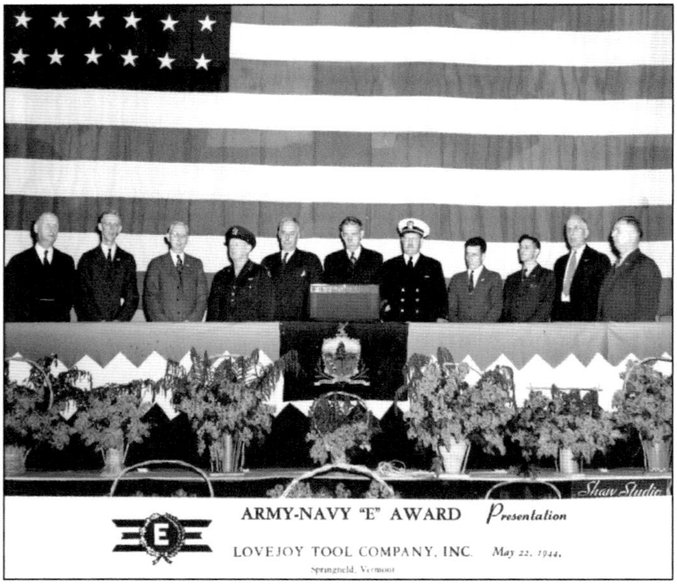


About 68 men and women at Lovejoy Tool Company received the Army-Navy E Award on May 22, 1944, in a ceremony at the Community House. This award was given in recognition of the company's fine record for production of tools used in the war effort. All four companies (J&L, Fellows, Bryant, and Lovejoy) and the foundry were recipients of this award. (STL.)

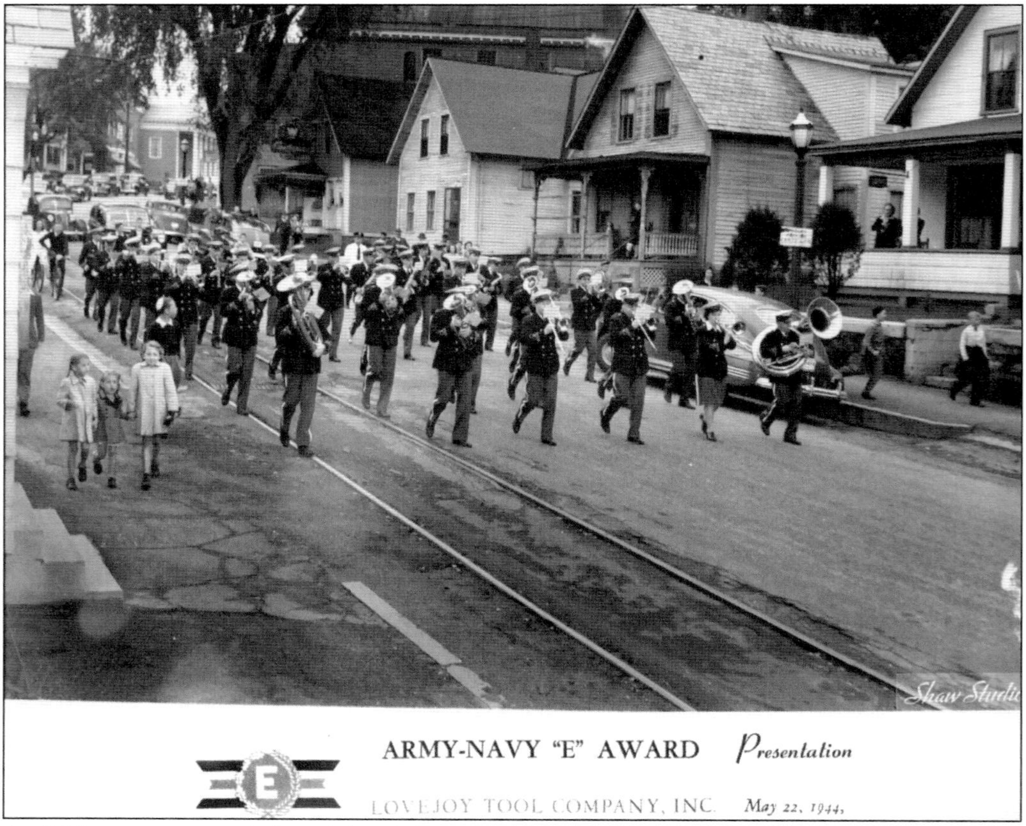


At 5:00 p.m., as part of the of the E Award celebration, the Gear Shaper Victory Band, with Barbara Giddings as drum majorette, marched from Fellows to the post office, where they gave a concert until 6:00 p.m. They then entered the Community House and took part in the ceremony. Thomas Chadwick, chairman of the shop committee, accepted the pins on behalf of the employees. (Lovejoy Tool Company.)

# *Six*

# North Springfield

Between 1881 and 1883, Louisa Griswold Field wrote the history of North Springfield for the Hubbard and Dartt book *History of the Town of Springfield, Vermont*. This seems to be the best history available. The earliest date she mentions is 1772, when John Griswold was in North Springfield. This was 11 years after the town of Springfield was incorporated.

North Springfield was its own village and quite self-sufficient. In the 1869 Beers *Atlas of Windsor County, Vermont*, the North Springfield inset on the map of Springfield shows a school, gristmill, sawmill, tin shop, two blacksmith shops, three stores (one containing the post office), and one church on Main Street. There were two more churches just off of Main Street. Businesses of some consequence in town were the cheese factory, the box factory, and the Boynton Sawmill.

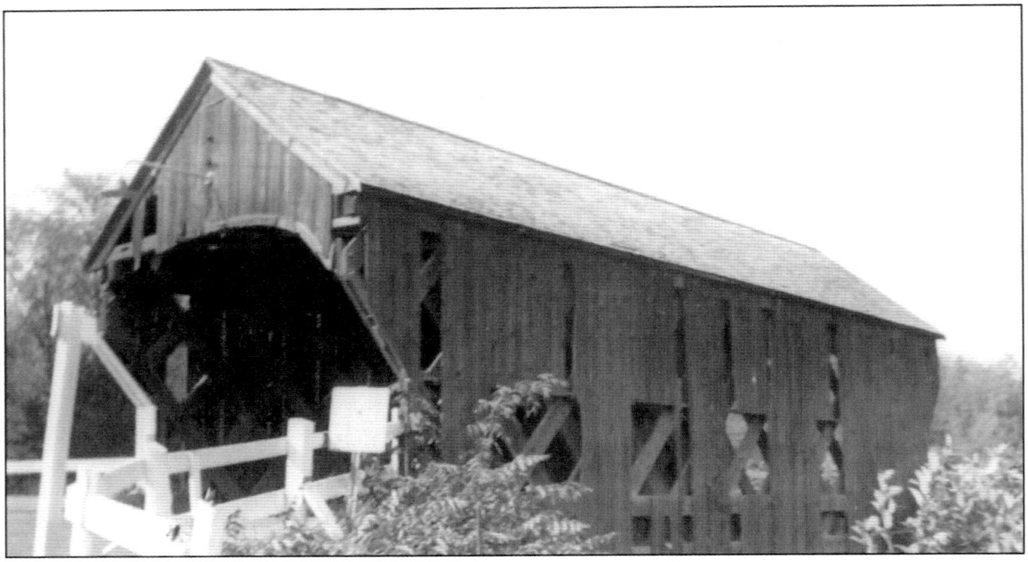

This was the bridge that went over the brook by the cider mill on County Road. It was moved when Route 106 was redirected and County Road was turned into the access road to the Industrial Park in North Springfield. The bridge was moved next to the relocated Eureka schoolhouse on Route 10 entering Springfield. (STL.)

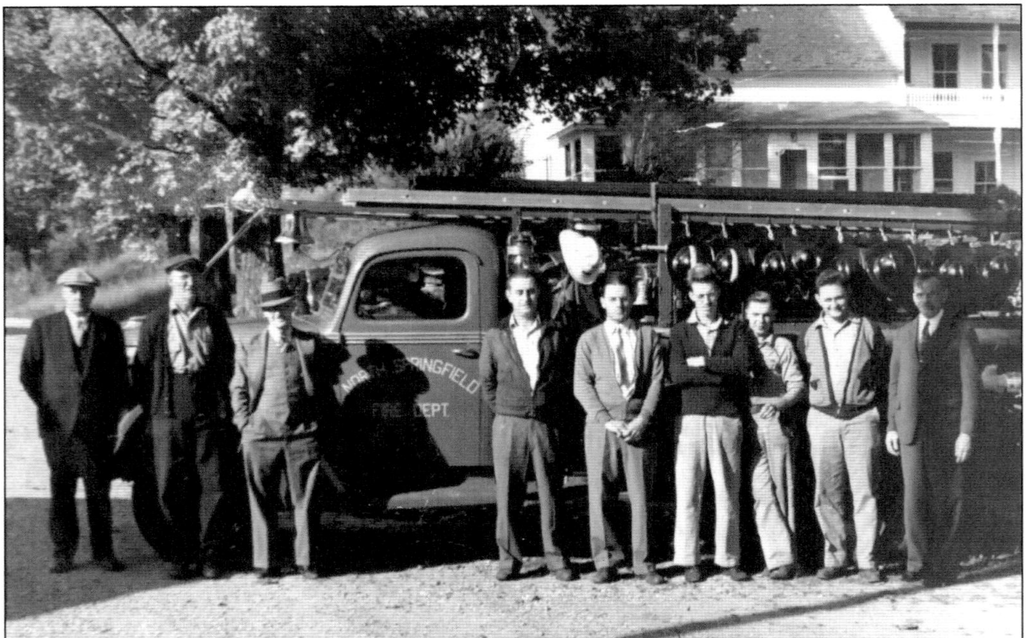


North Springfield had its own fire department. This department was critical in saving the IOOF Building in Springfield during the Stiles fire in 1934. Included in this 1943 photograph are, from left to right, Ralph Austin, Chief Charles A. Hannah, Fred Gibson, George Thompson, Raymond Hastings, C. Edwin Hannah, Dan Dyshlevich, Tommy Thompson, and Rev. E. Thornton. (Kaye Peck.)

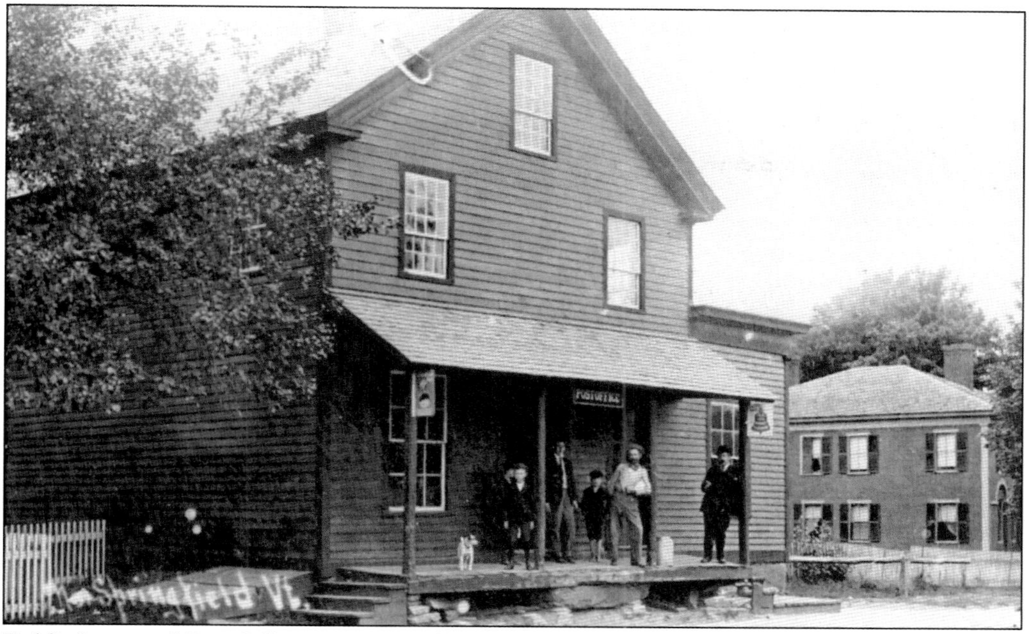


Field's Store and Post Office was originally built by James Litchfield as a home for his son Otis. It was converted to a store in 1831, and Abner Field became the first postmaster in 1832. His son Fred G. ran the store for many years, but it burned down in 1915. (WGS.)

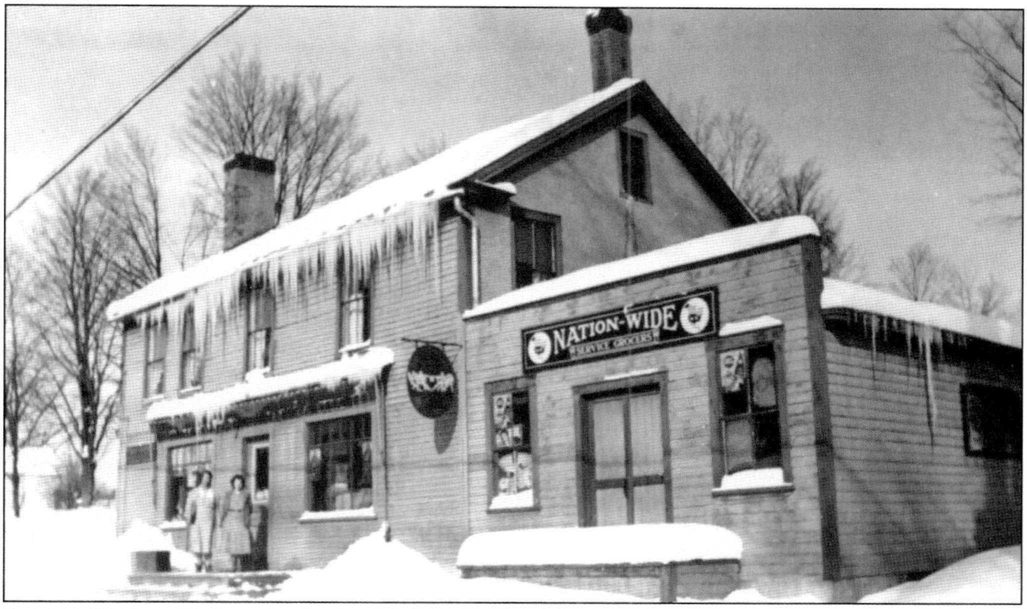


In 1916, William Fuller, who had worked for Fred Field, opened a new store on the corner of School and Main Streets (now known as the Main Street Market). In 1925, Fuller would sell the store to E.J. Chapman. The store has been in operation, with various owners, since that time. In the photograph are Marjorie Chapman (right) and Sybil Spafford. (Main Street Market.)

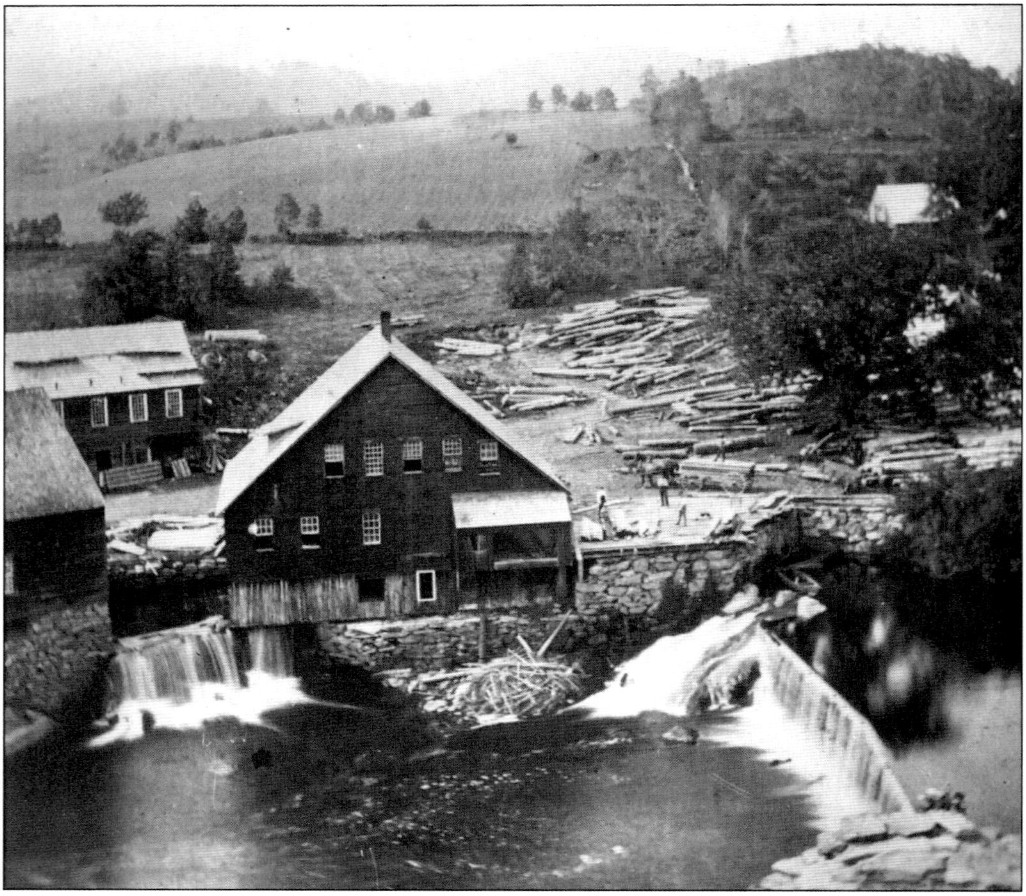

Boynton's Sawmill was built between 1870 and 1871 on the site of an old mill that had been carried away by the 1869 flood. Durant J. Boynton owned both the sawmill and waterpower. The mill could produce 500,000 feet of lumber a year in addition to shingles and lathes. (WGS.)

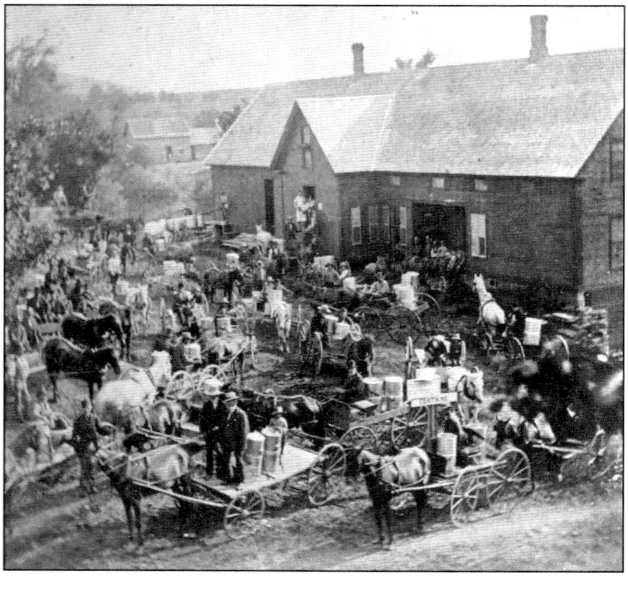

C.A. Leland and F.G. Field built the cheese factory on the corner of Maple and Main Streets in 1877. Annually, it produced 30,000 to 60,000 pounds of cheese from the milk of 150 to 300 cows, but it had the capacity for up to 400 cows. The building is now a residence. (WGS.)

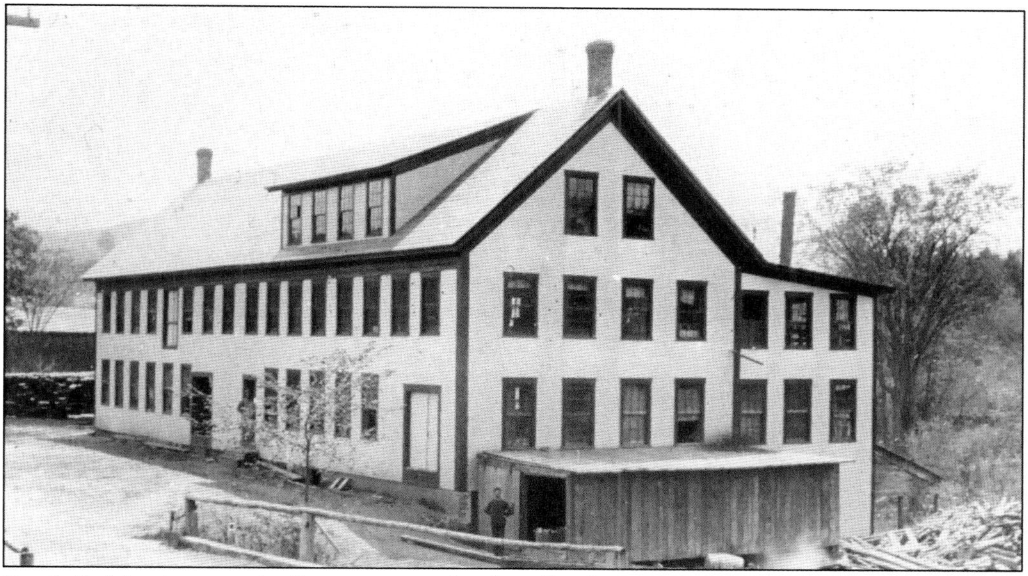

Frank D. Martin started off turning wooden handles for wire utensils. In 1879, he switched to making boxes. In September 1884, his building burned to the ground. Martin rebuilt, constructing this three-story edifice where he employed 26 workers making both plain and fancy wooden boxes, including jewelry boxes and silver cases. However, it lasted less than 10 years, burning down again in 1893. (WGS.)

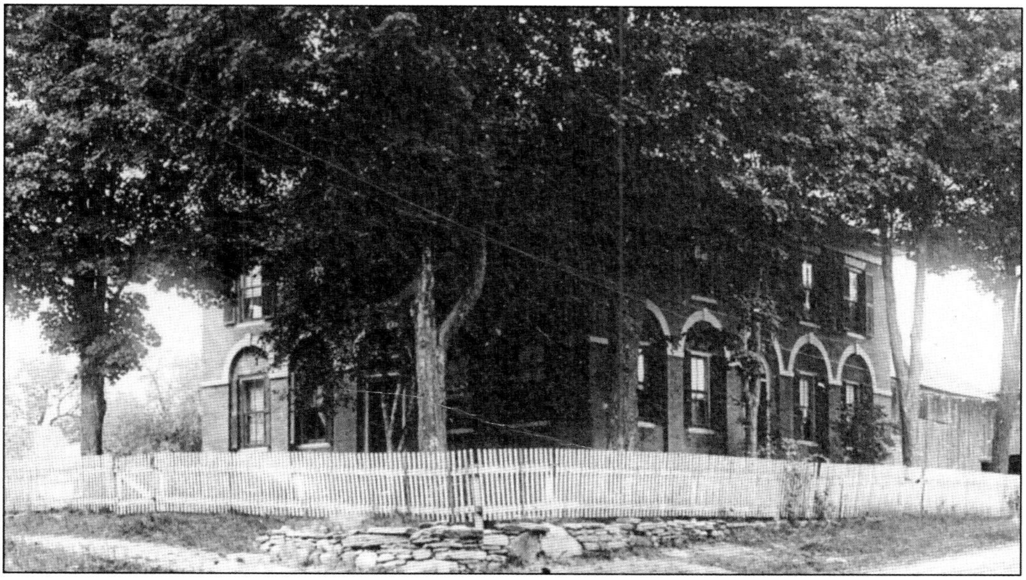

Erected about 1800, the brick facing on two sides of this building was added in 1813. It was originally smaller and housed a tavern, while the entire upstairs was a dance floor. Later, the house was made square by adding six wood-framed rooms to the west side, centering the front door, and partitioning the upstairs into rooms. It was known as the Boynton or Fuller House and is now the Brookwood Nursing Home. (WGS.)

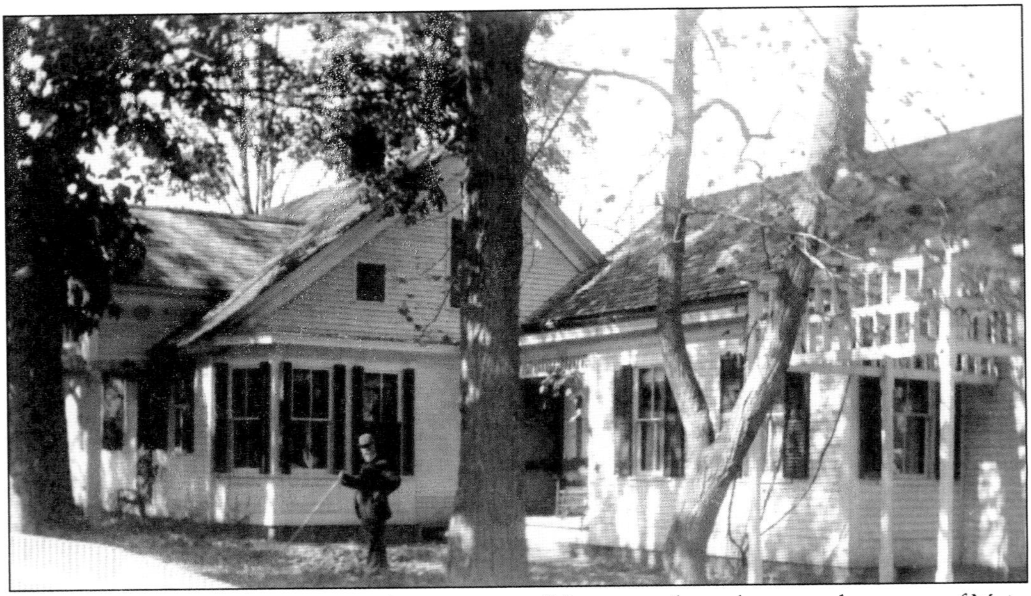

The Hannah house on Main Street began as a small four-room frame house at the corner of Main and School Streets. It was moved to make room for the brick house (Brookwood) on the corner and was then purchased by Abner Field in 1831. An addition was adjoined on the west end in 1845, and later another was added on the east end. (STL.)

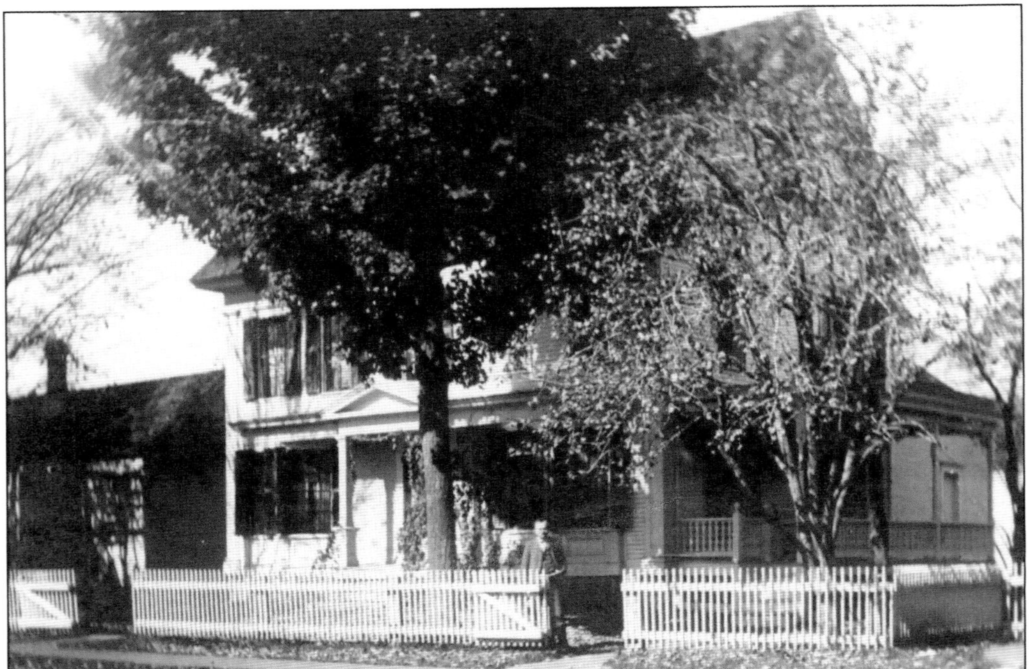

Prior to marrying Anna Tarbell in 1872, Fred G. Field (Abner's son) built this house in 1871. Although it is hard to tell now, it was once sandwiched between Abner's house and the Field Store. If looking at the lawn to the right of the house today, the old foundation of the store can be seen. It is amazing that the house did not burn down along with the store. (STL.)

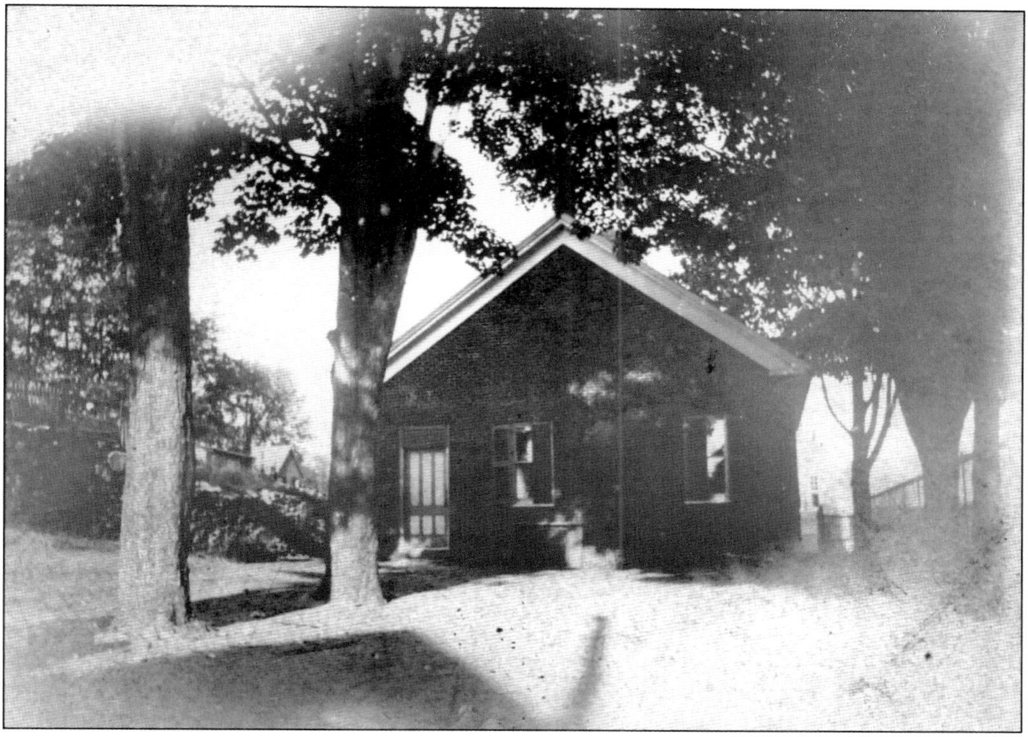

In 1907, a special town meeting was called to consider adding a second room to this schoolhouse. However, it was decided to build a new four-room building at a cost of $10,000 instead. The contract was let in 1908, and the school was completed in 1909. (SAB.)

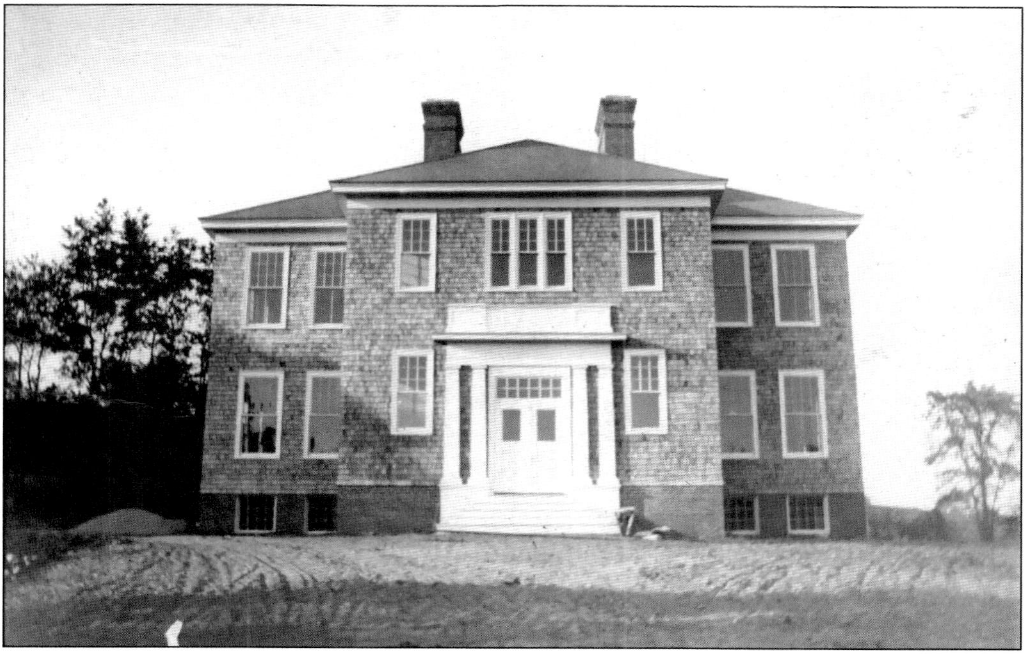

By 1929, this school, completed in 1909, was already too small. However, it would not be until 1938 that a new addition would be constructed, providing two more classrooms and an unfinished space to use for a hall or future classrooms. The school was closed in 1984. (SAB.)

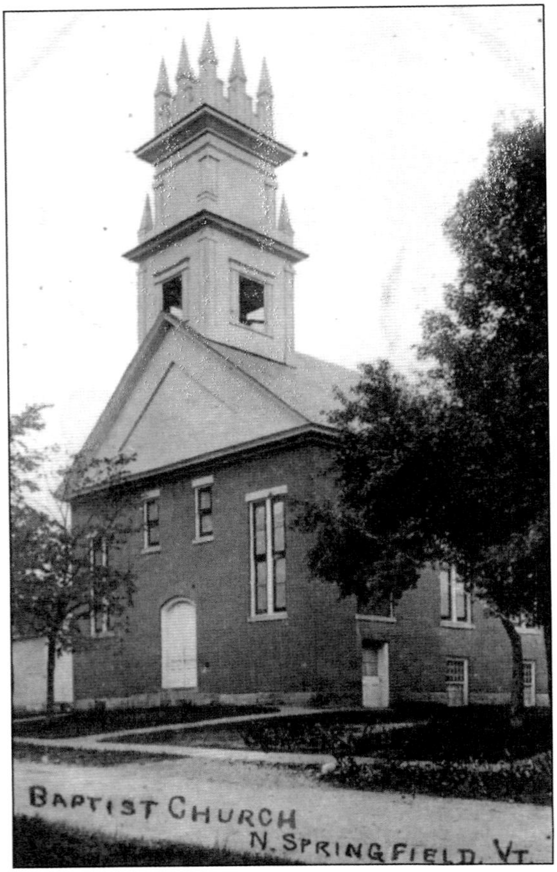


The North Springfield Baptist Church was built in 1835. The first school had been erected on this spot in 1811, but it burned after a few years. The parsonage next door was not built until 1870. (STL.)

The Advent Church was originally a Congregational church in Perkinsville. It was moved to its present site on Church Street around 1856. The horse sheds are visible in the back. (STL.)

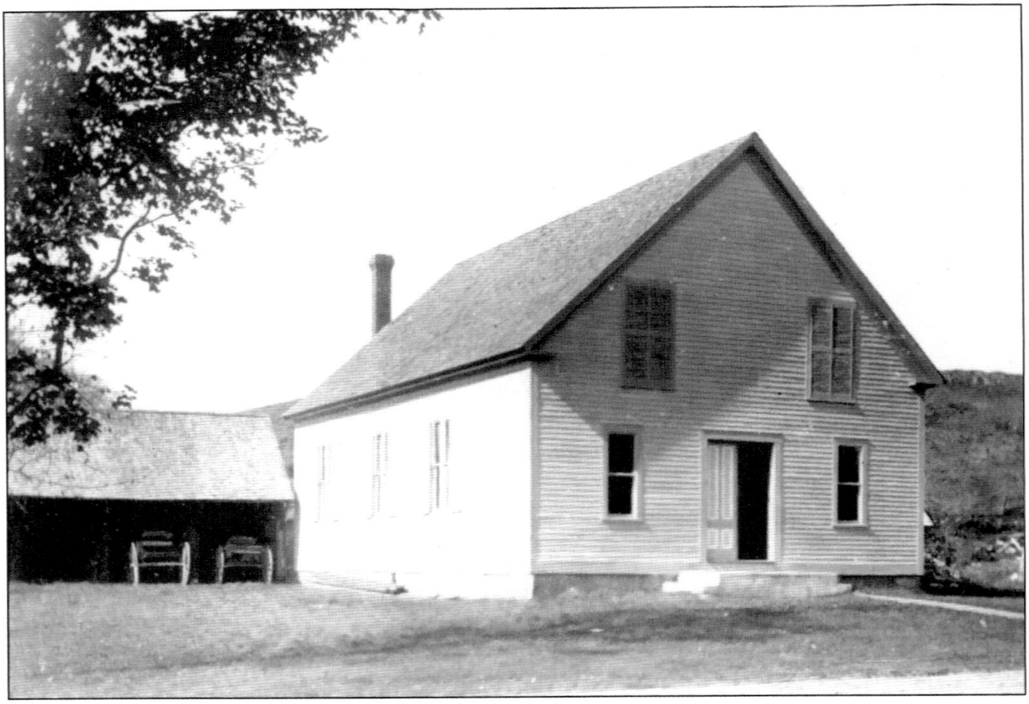

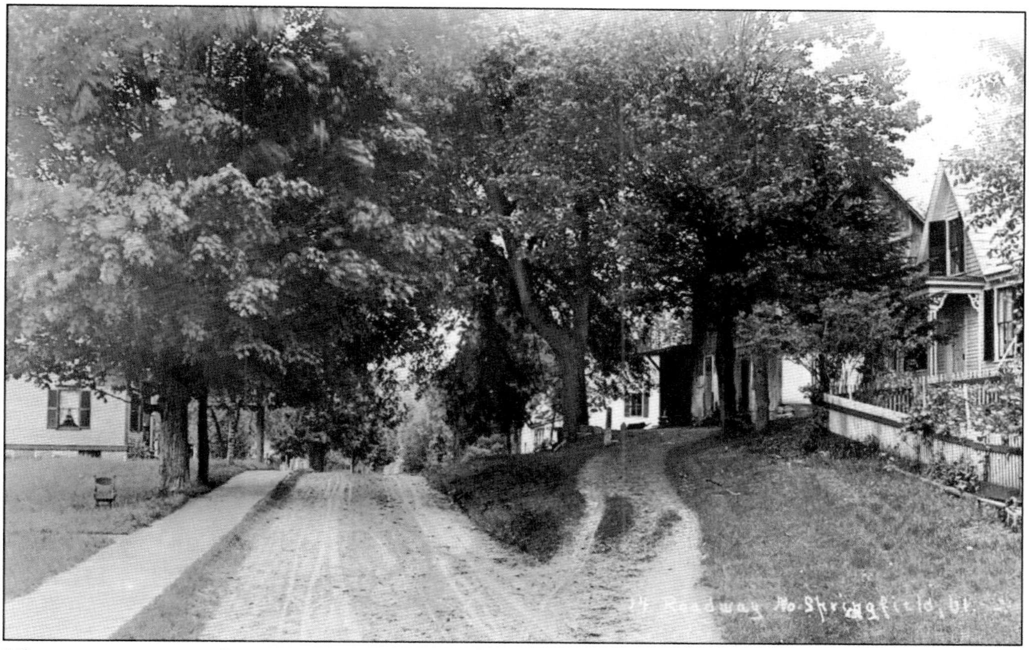


This image was taken on Main Street, looking west towards the center of town. The Boynton house (right) is the first on the right on Clark Street when heading into the Clark Development. (STL.)

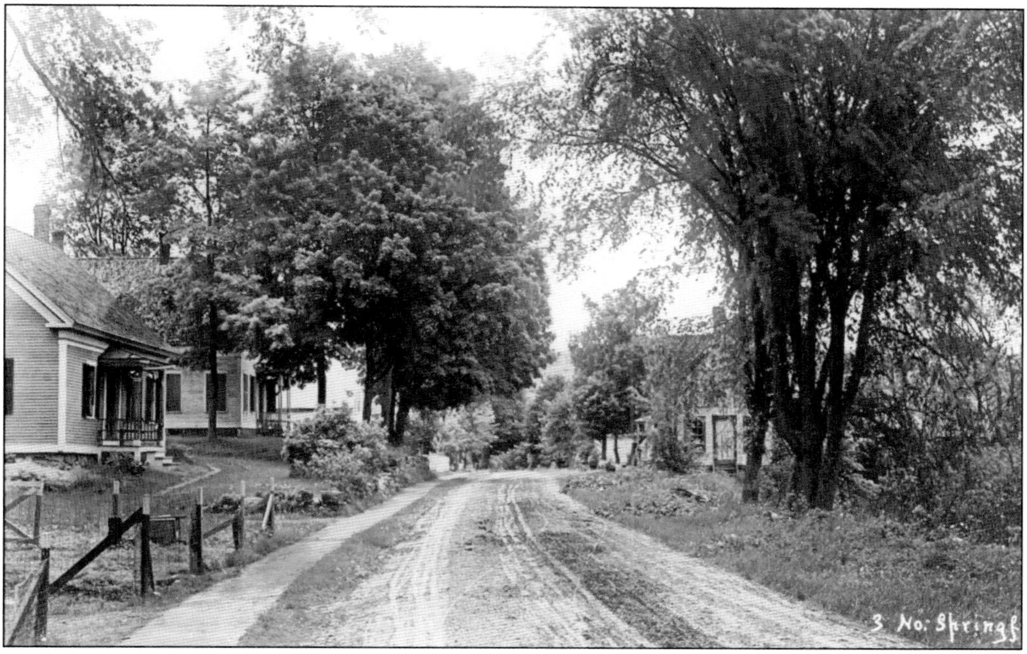


This photograph was taken facing the opposite direction, traveling east. Just out of sight on the left is the North Springfield Baptist Church. (STL.)

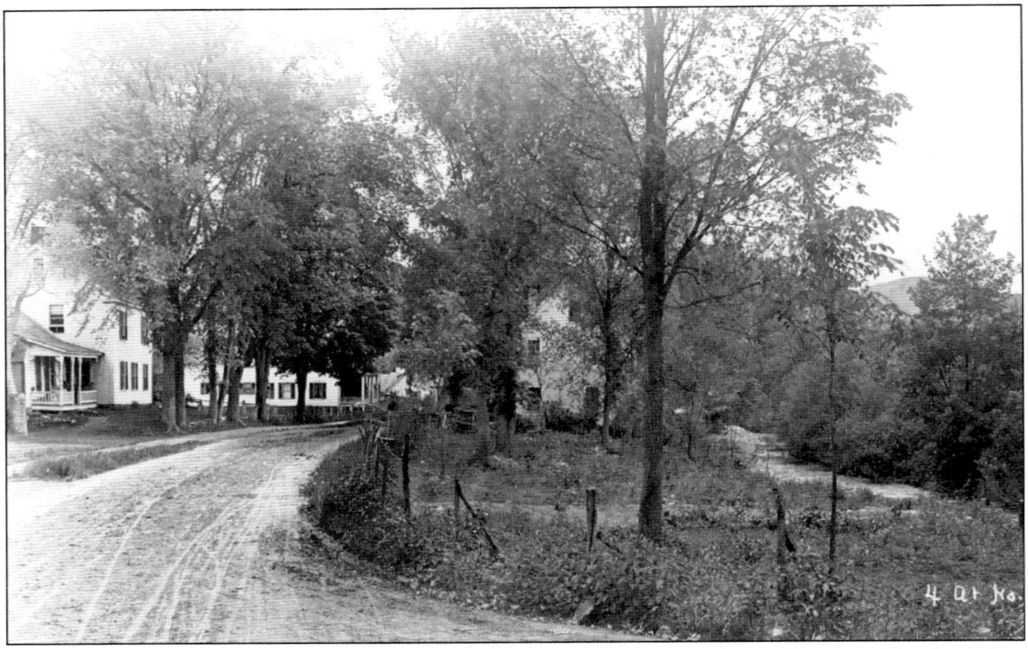

This is a view from the corner of Church and Main Streets across from the store. On the left is the Martin House. It looks a little different, but the two windows over three windows on the side of the house facing the road are visible. (STL.)

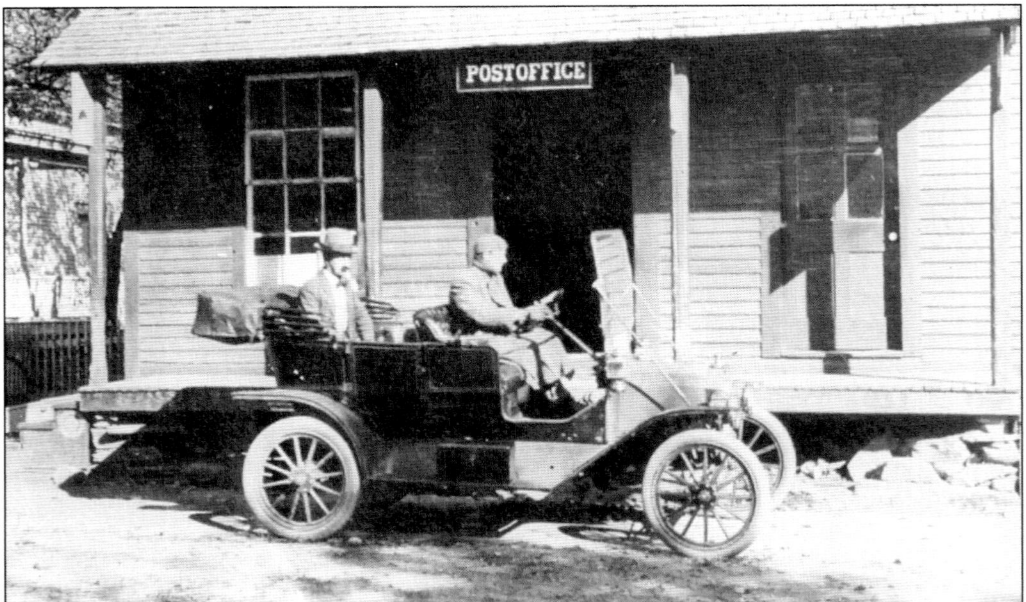


Here is William Fuller in front of Field's Store in his 1909 Maxwell. John Bemis is the driver. It may have been the first car in North Springfield. (WGS.)

# *Seven*

# Leisure and Sports

From the time the town hall was built in 1858, plays, minstrel shows, lectures, dances, and even movies were performed or shown there.

At the turn of the 20th century, leisure in Springfield included field days, Fourth of July celebrations (complete with a big parade), and the annual fair generally held in September, with balloon ascensions and horse racing as the main attractions.

In 1925, the Blue Sox, a traveling baseball team, was formed and continued to entertain crowds through 1944. Their games were always a big attraction after the Fourth of July parade. At the same time, there were several other teams in town. Some were neighborhood teams, and others were formed by the various shops in town. There were always high school athletic programs to attend as well.

Springfield has its own Olympic gold medal winner, Albert Gutterson. In 1911, prior to the 1912 Olympics, while competing at a University of Vermont (UVM) track meet with the University of Maine, Gutterson scored 33 of Vermont's 42.5 points. He entered seven events and won six of them. They were the 100-yard dash, 200-yard dash, high jump, broad jump, discus, and 220-yard hurdles. He placed second in the high hurdles.

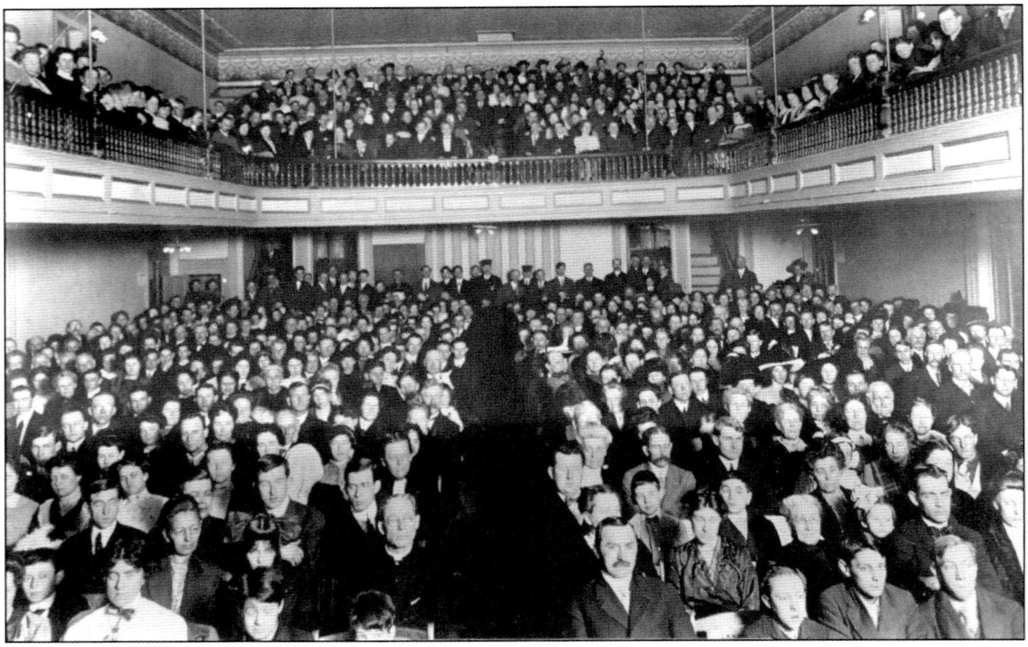

The town hall was originally a two-story, one-room auditorium with a balcony. It was the social center of the town until 1919 when the Community House opened. Town meetings, dances, and concerts were held here. Traveling minstrel shows and bands would also perform here. John Philip Sousa's band even came to town. There are at least 300 people in this photograph that was taken during Dodge's White Store Concert in 1913. (STL.)

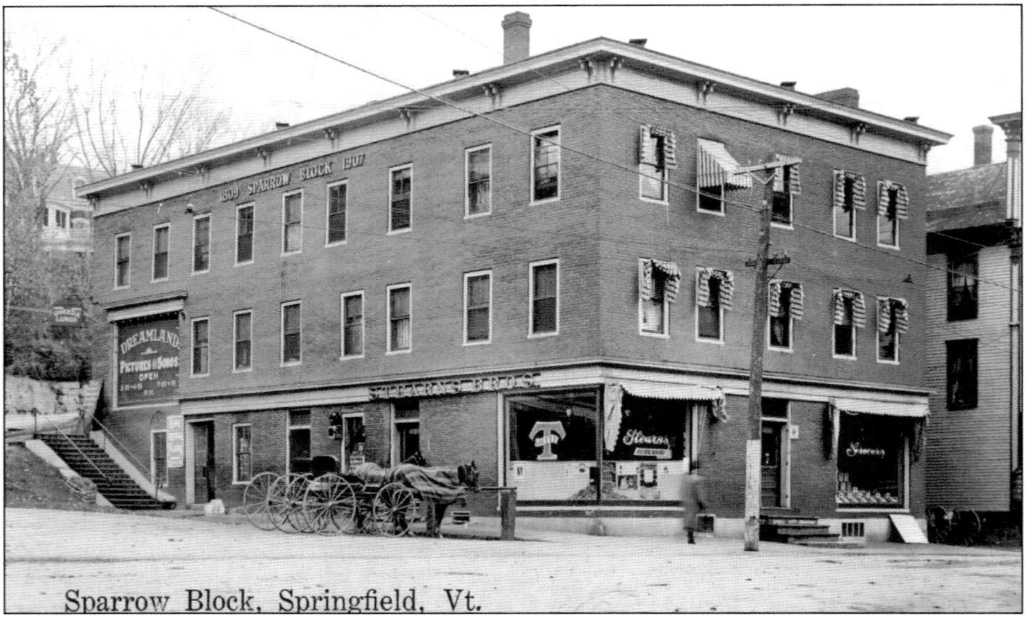


In July 1909, Dreamland opened to show silent movies with musical accompaniment. It was located on the second floor of the Sparrow Block, and the entrance was from Summer Hill. It could seat about 150 people. By 1915, Dreamland had been remodeled and updated as the Ideal Theater. Patrons entered from the square. By this time, it had a balcony and could seat 420 people. (HSP.)

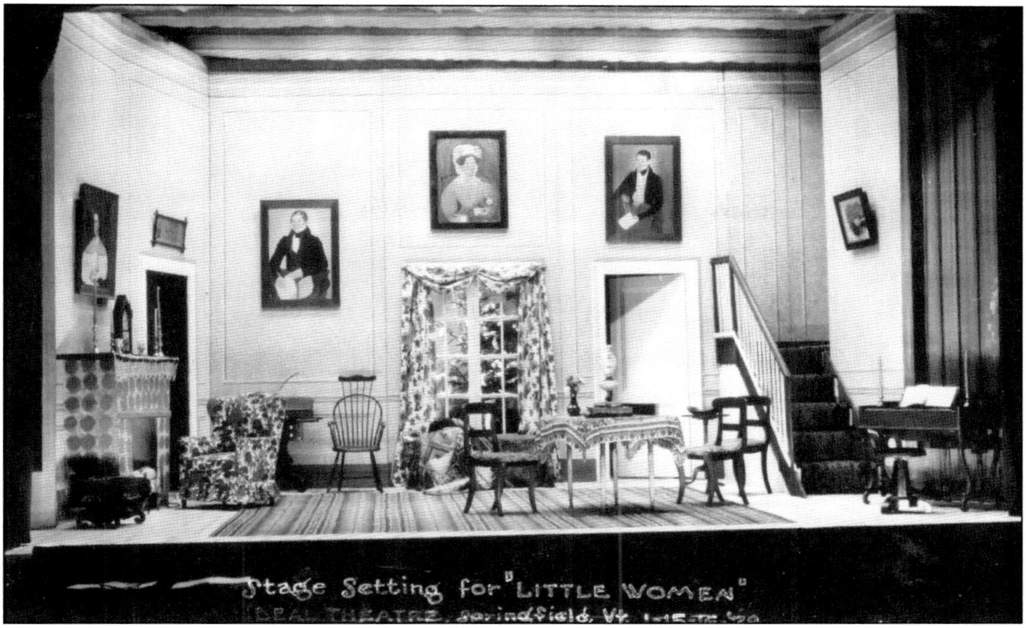


Although normally a movie theater, on January 15 and 16, 1929, the Women's Association of the Congregational Church presented the four-act play *Little Women* at the Ideal Theater. All seats were reserved at 50¢ and 75¢. There would be a concert by the Ideal Theater Orchestra (piano, flute, violin, clarinet, and drums) at 7:45 p.m., and the play would begin at 8:30 p.m. (STL.)

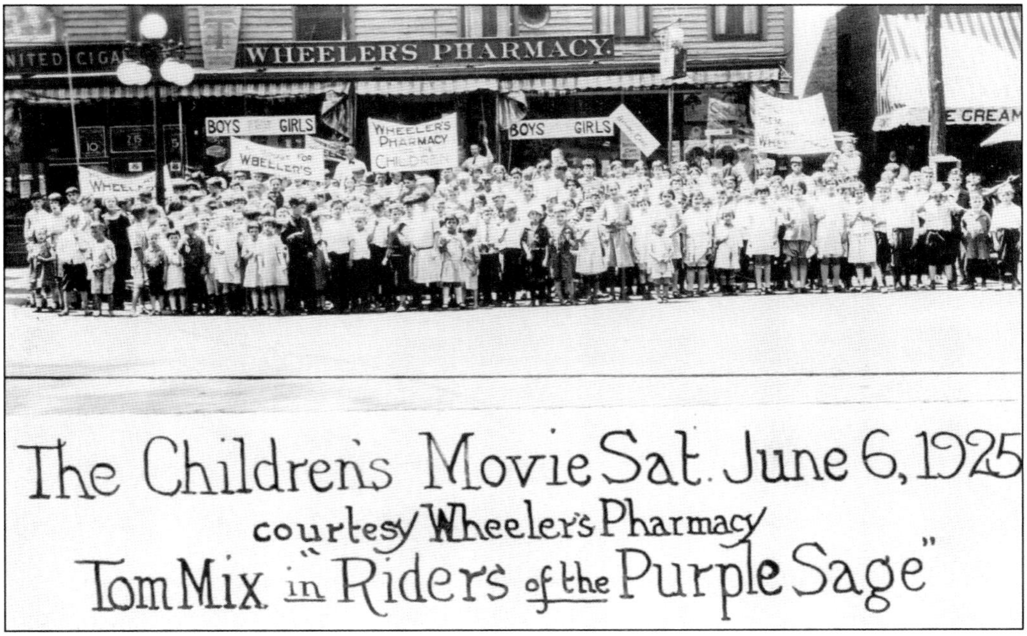


A Tom Mix movie was screened for the younger crowd. This cowboy was very popular, and so was Wheeler's Pharmacy on June 6, 1925 when over 100 children received a ticket to the Saturday matinee at the Ideal Theater across the street. (STL.)

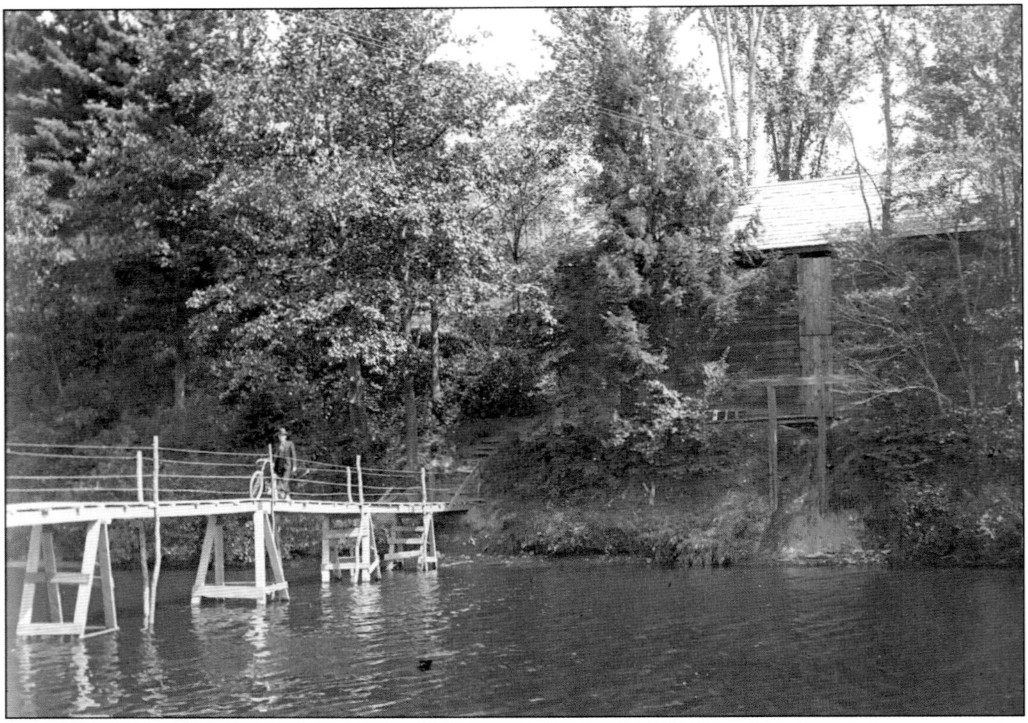

Until 1906, the Falls Bridge and the bridge at North Springfield were the closest routes over the river to Riverside. A temporary bridge was erected each year near the icehouse for fairgoers. This was important because the fair could attract between 3,000 and 9,000 people. (WGS.)

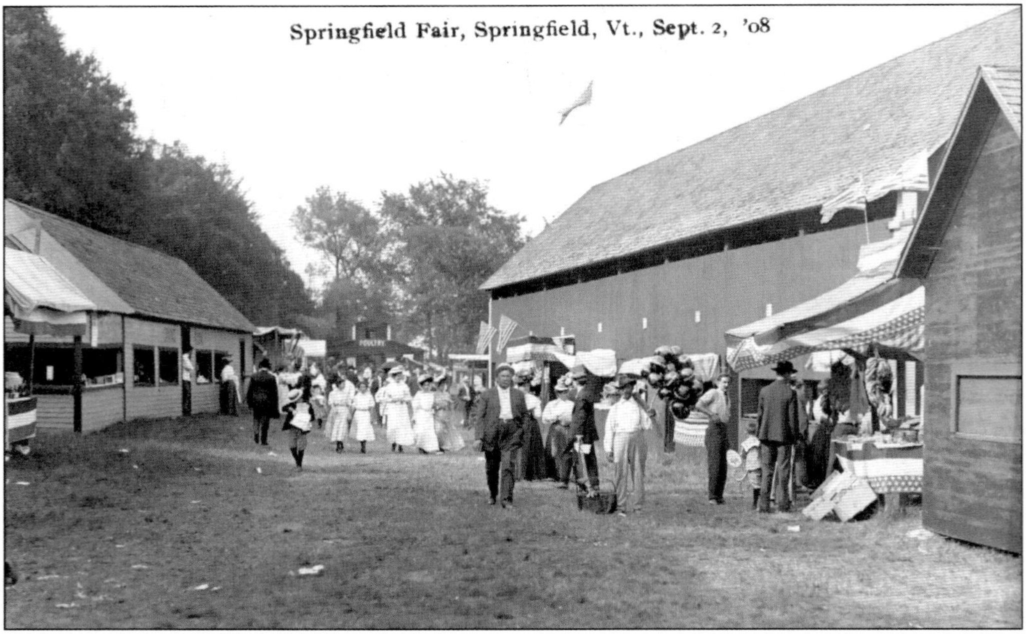


There was a midway with a floral hall, animal judging, and games and contests with ribbons or prizes. Displays and demonstrations of the latest farm and home improvements abounded. Held for two days, the hardest part was getting through it with no rain. Many times, the whole affair had to be postponed for a day or two. (STL.)

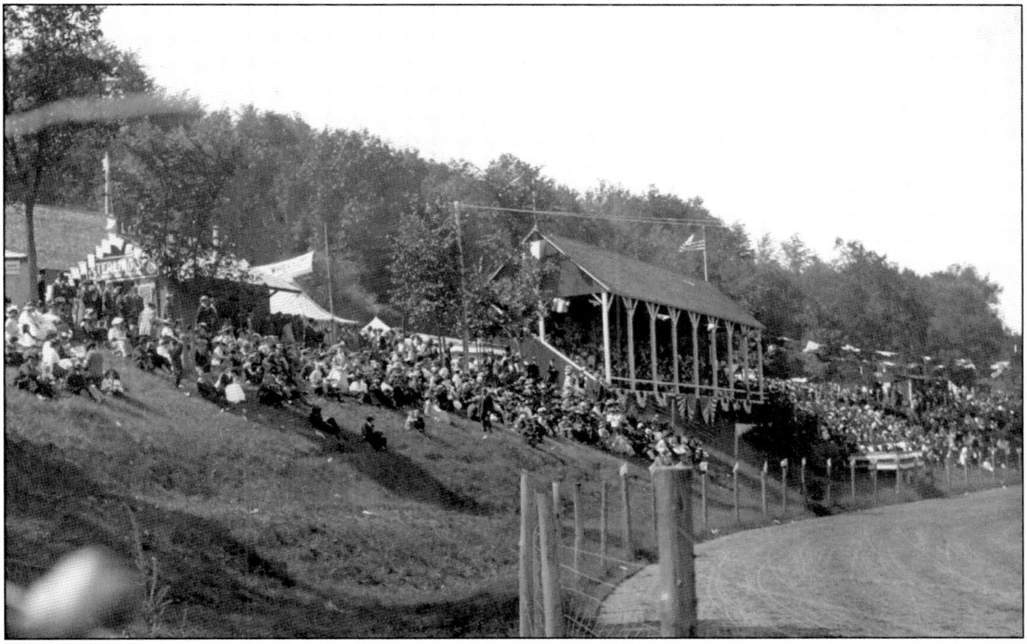

The grandstands built in 1896 would fill early. Others would bring out their blankets or sit in their buggies. In 1899, it was advertised that an automobile would be on display. In 1900, the fair advertised a five-mile automobile race. On the first day, the cars would not start. On the second day, the drivers had second thoughts. The balloonist was more exciting. The balloon was released, and the performer jumped out with a parachute. In 1894, the balloon caught fire and the event was cancelled. In 1896, the ascender missed his mark and landed in the river. In 1908, he landed in a tree on Park Street Hill. In 1926, a helper became entangled in the ropes. The balloon went up about 50 feet before he fell to the ground and had to be hospitalized. (Both, STL.)

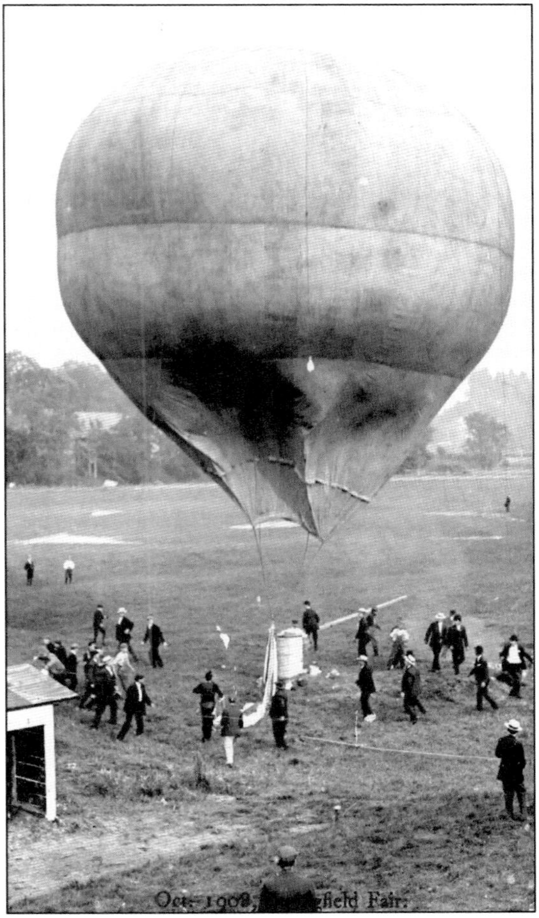

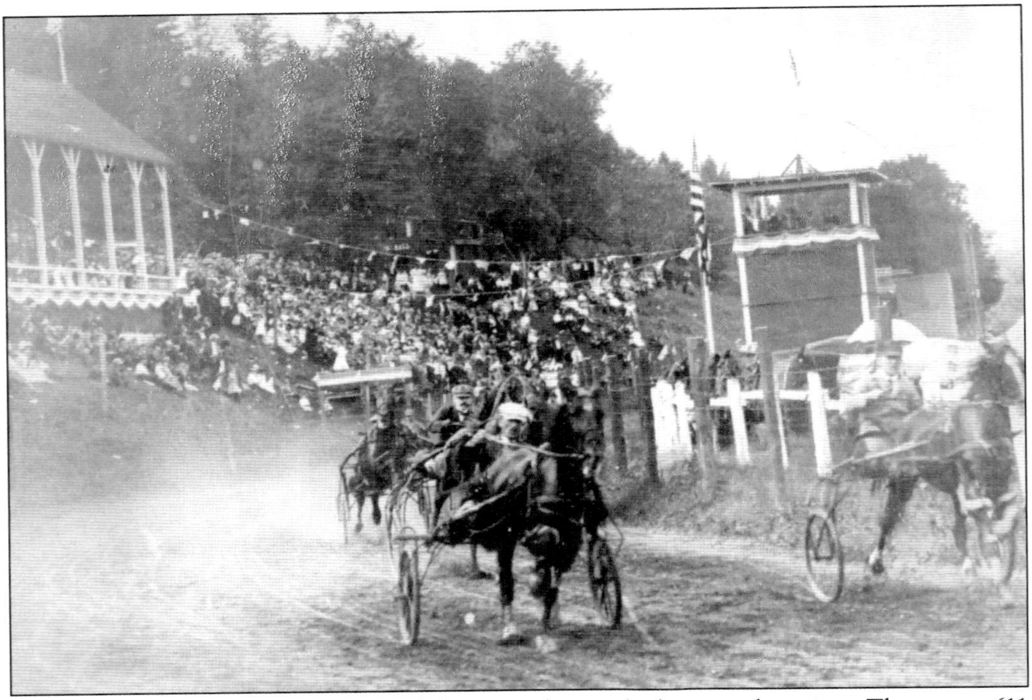

Times were changing, and the automobile was replacing the horse and carriage. There were 611 automobiles parked inside the track in 1920. In 1930, the fair came to an end. It might have been because of the mobility of the car or because of the Depression. With an agreement that the town pay the association's debt of $5,932.19, the Springfield Fair Association transferred the property to the town. (STL.)

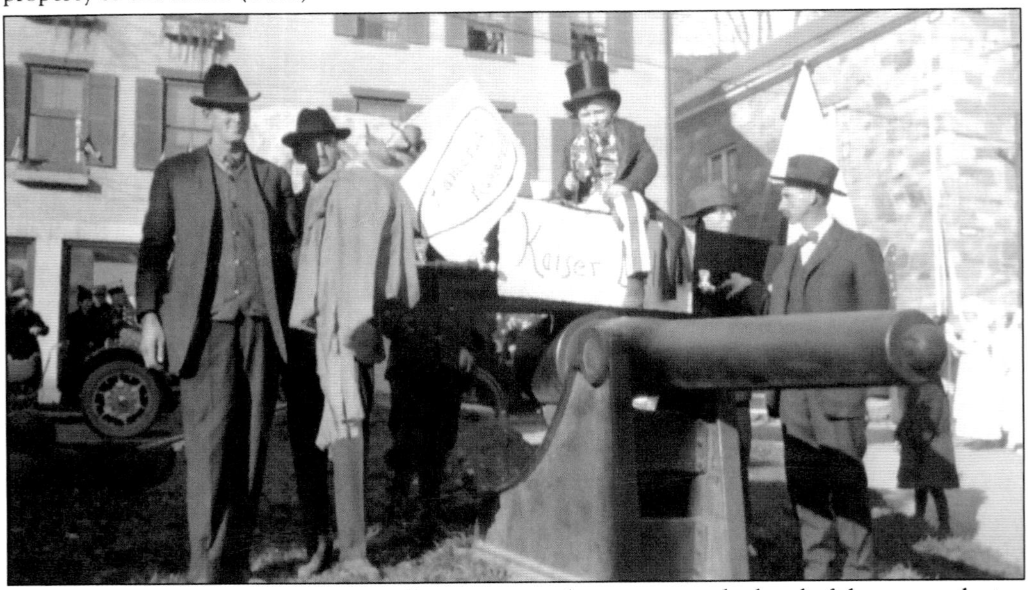


In this image from 1918, Proctor Lovell is sitting on the cannon at the head of the square during the celebration of Armistice Day. Although much discussion has taken place about what vintage this cannon was, it is known that it was donated by the Women's Relief Corps (WRC) to the World War II scrap drive in October 1942. The WRC received it from W.H.H. Slack some 50 years earlier. (STL.)

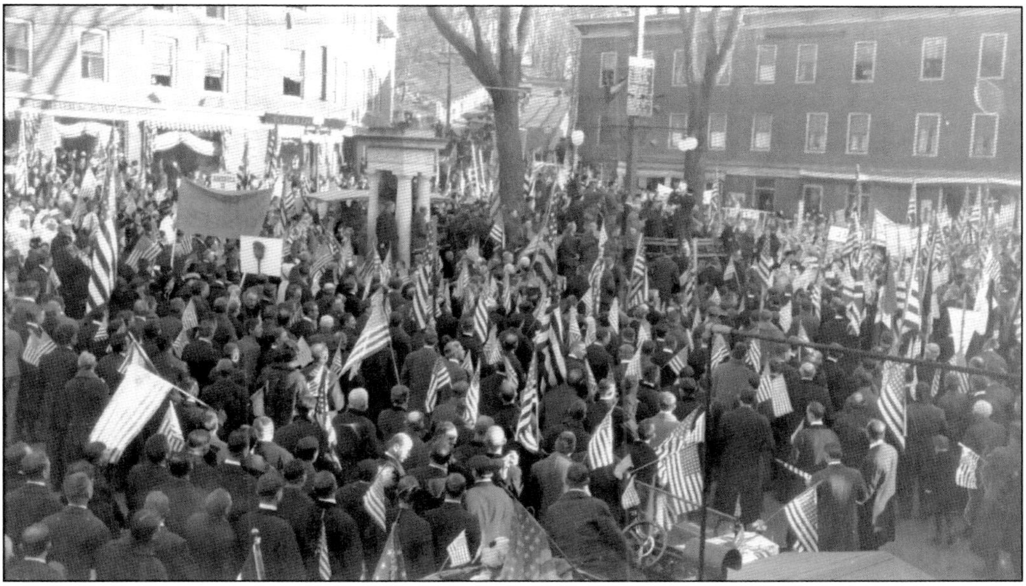

The biggest parades in Springfield had been on the Fourth of July. Later, the alumni parade would be the biggest, but this photograph is of the 1918 Armistice Day Parade. Groups of every sort came out for this event. Everyone was so happy that World War I was over. (STL.)

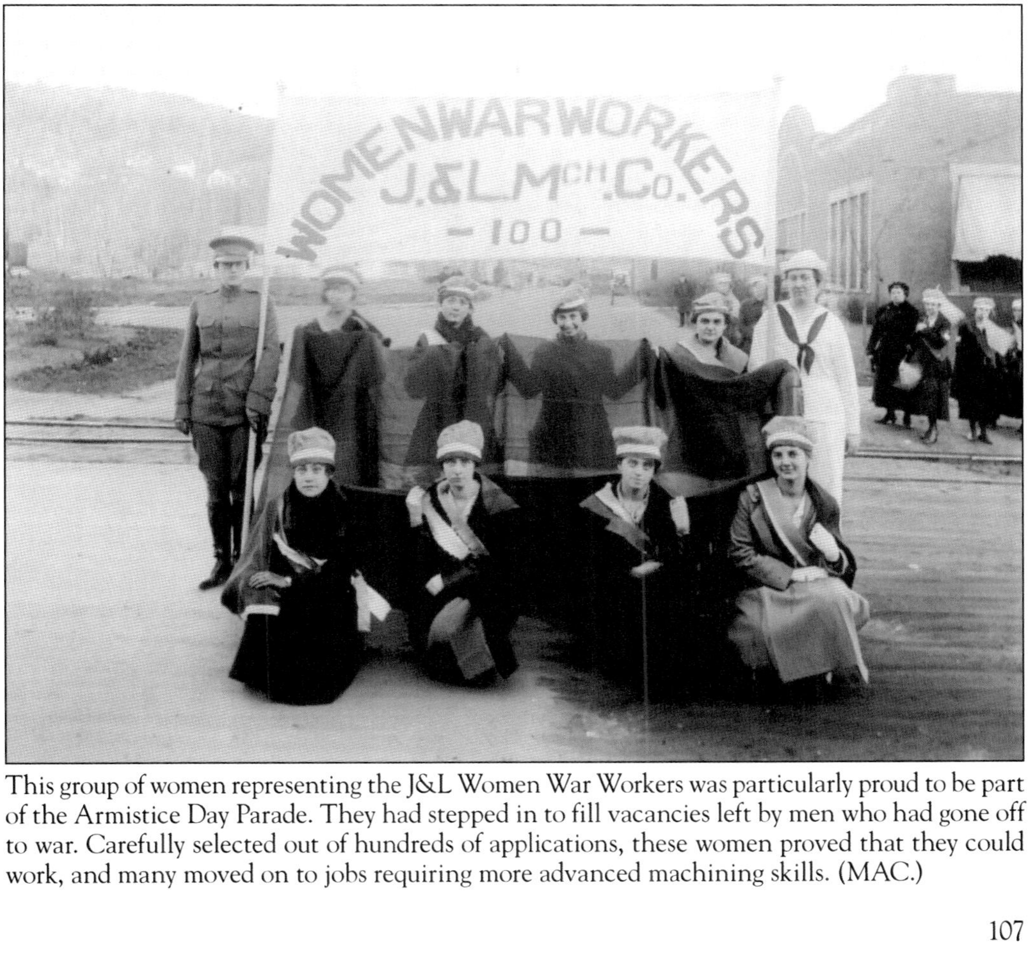


This group of women representing the J&L Women War Workers was particularly proud to be part of the Armistice Day Parade. They had stepped in to fill vacancies left by men who had gone off to war. Carefully selected out of hundreds of applications, these women proved that they could work, and many moved on to jobs requiring more advanced machining skills. (MAC.)

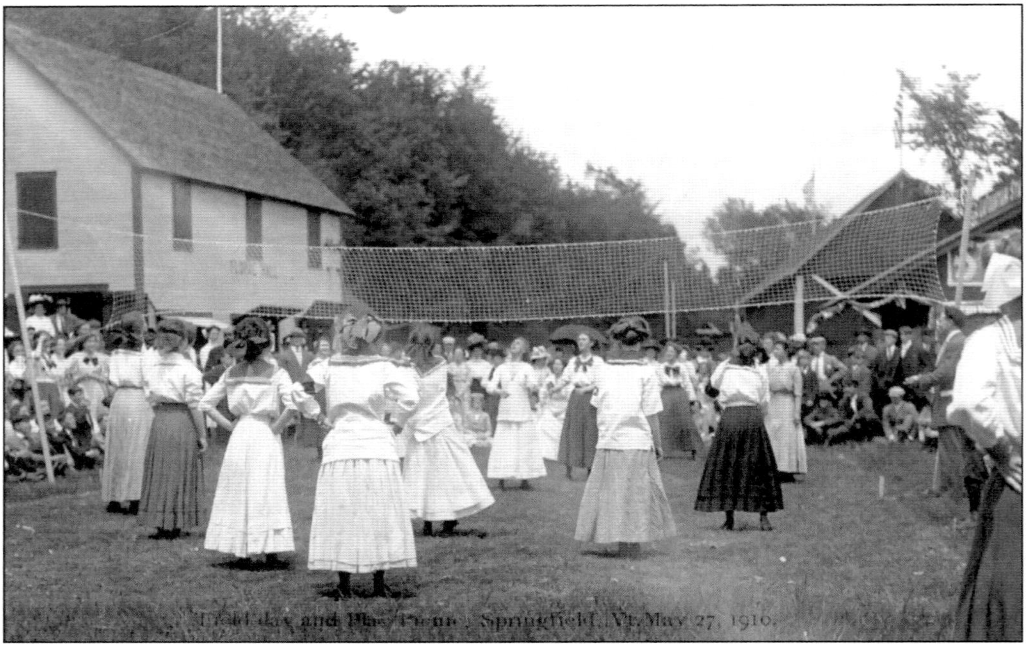


There were also family play days, often sponsored by the Young Men's Christian Union (YMCU). These photographs are from 1910. In this image, the ladies are playing a volleyball game in their long skirts. (STL.)

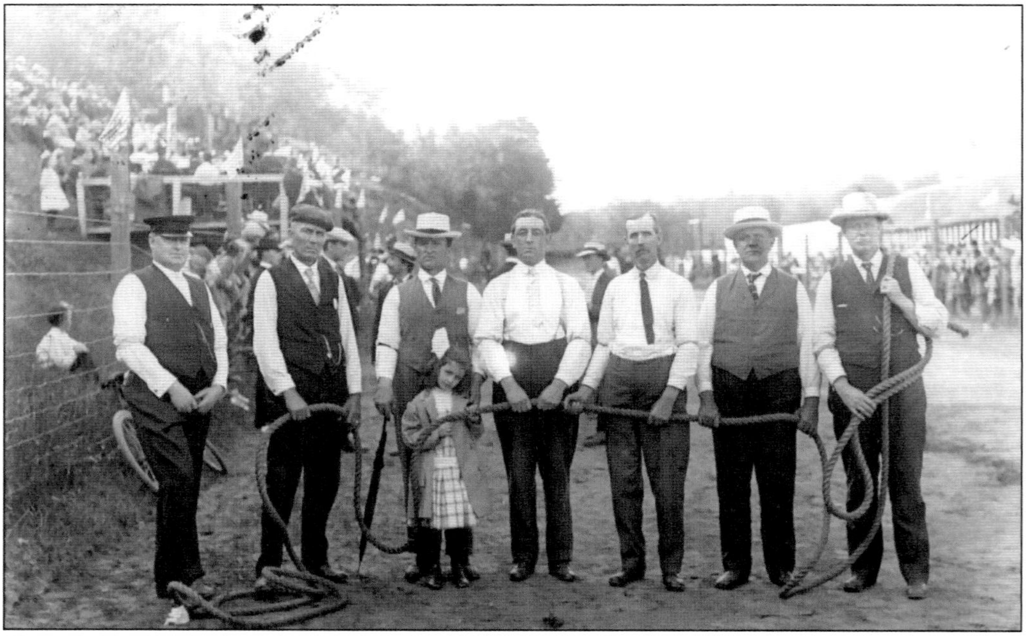

The winning businessmen's tug-of-war team was, from left to right: Dr. L.H. Gillette, police chief Ed Purdy, Frank Gill and daughter Gail, Arthur Stearns, Walter Stearns, Franklin Barney, and George Leland. (STL.)

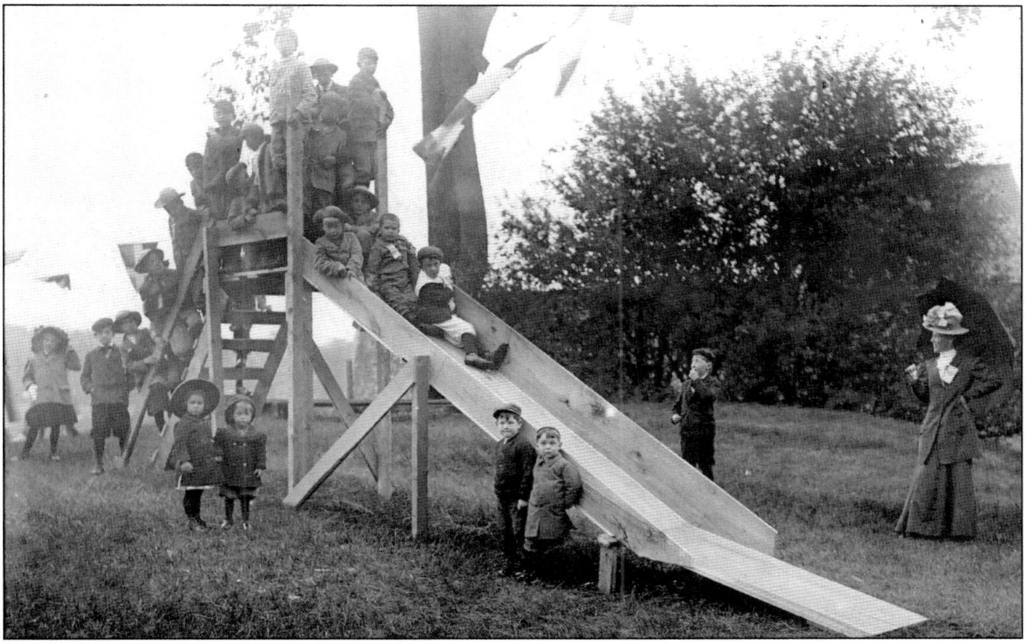

There was also entertainment for the little ones, including a large swing set, a sandbox, and this lovely wooden slide. (STL.)

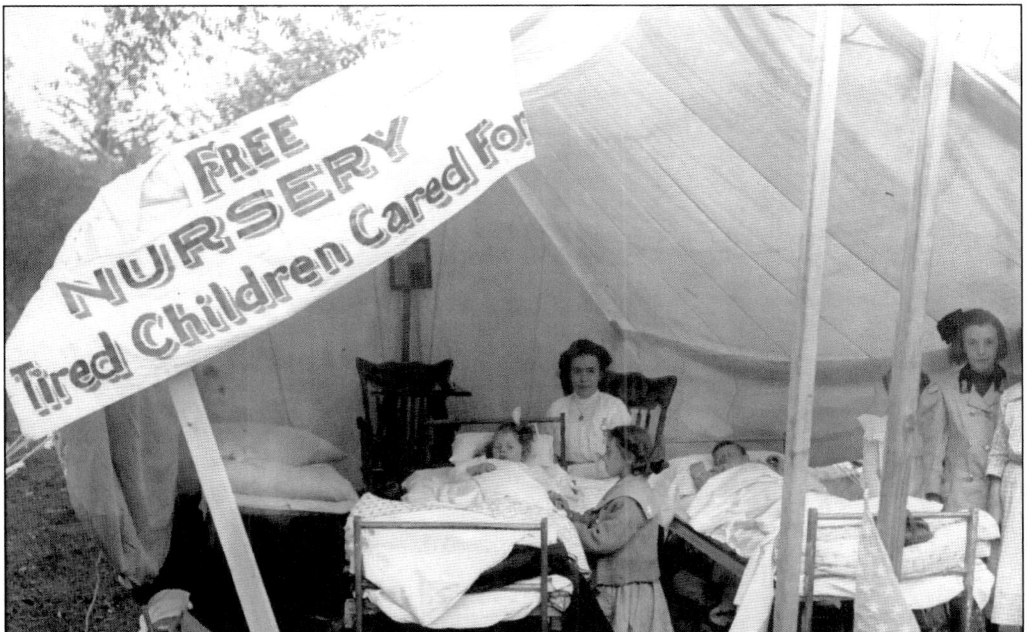


When the youngsters got overly tired, they could be brought to this free nursery where "tired children [were] cared for." It does not look like they expected too many tired children, unless there were other tents. (STL.)

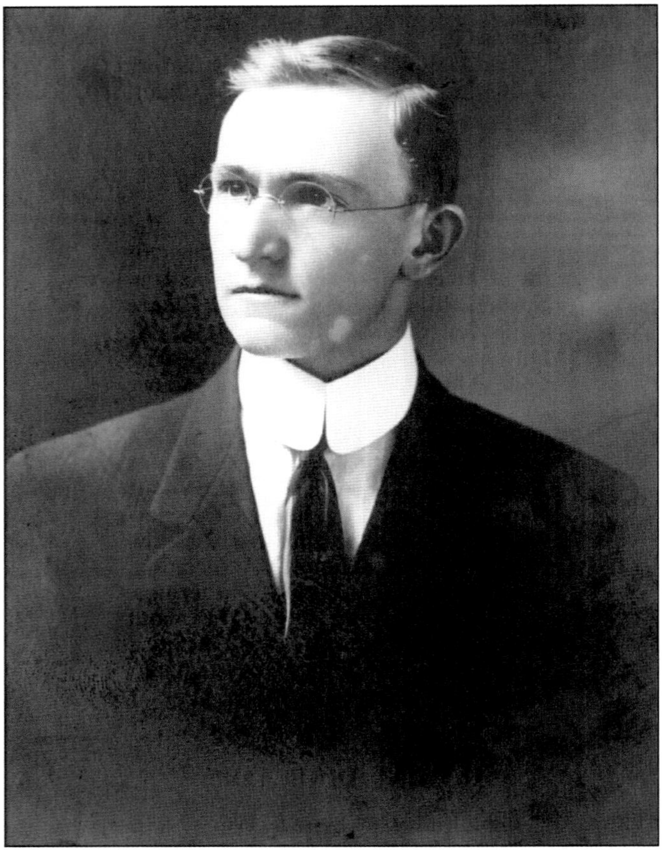

Albert Gutterson was a graduate of Springfield High School and the University of Vermont. He competed in track and field at the 1912 Stockholm Summer Olympics and won a gold medal in the running broad jump (24.93 feet). Gutterson Field House at UVM is named for him. He returned to Springfield and was president of Lovejoy Tool from 1950 to 1963. (Special Collections, Bailey-Howe Library, UVM.)

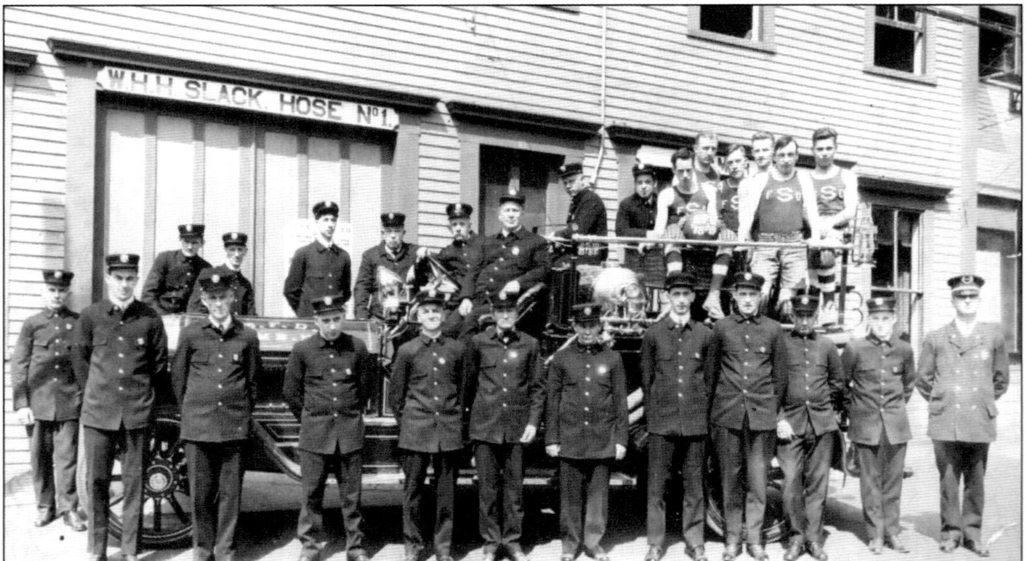


Taken around 1919, the basketball team photographed on the fire truck was most likely sponsored by the fire department, given the letters on their shirts. This was not unusual since, until the 1940s, there were leagues of all sorts in town. Some were shop or neighborhood teams that played locally among themselves, and others would travel and play teams from other towns. (STL.)

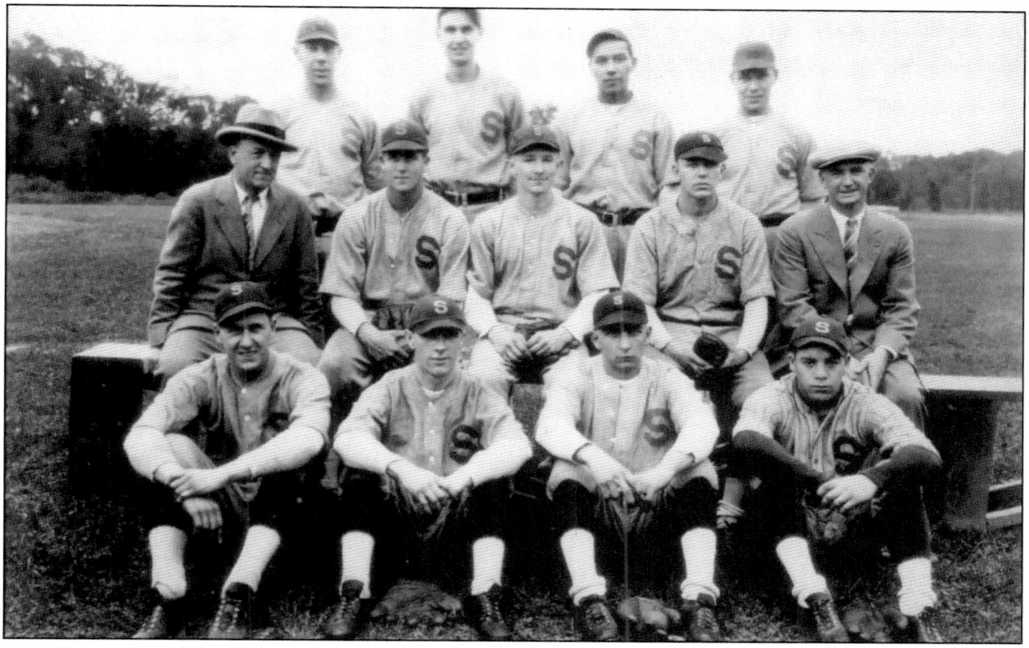

The Springfield Blue Sox were a traveling baseball team active from 1924 to 1944. After the war, people seemed to lose interest, and the team disbanded. The 1927 Blue Sox Team are, from left to right, (first row) D. Currie or Vanio Kangas, Clark Corliss, Henry Stone, and Carlton Raymond; (second row) Ernest Illingworth, Curt Dressel, Dan Mahoney, Heine Wilhelm, and Henry English; (third row) Ernest Schoenfeld, Simp Simonds, Tommy Owl, and Ernie Phillips. (MAC.)

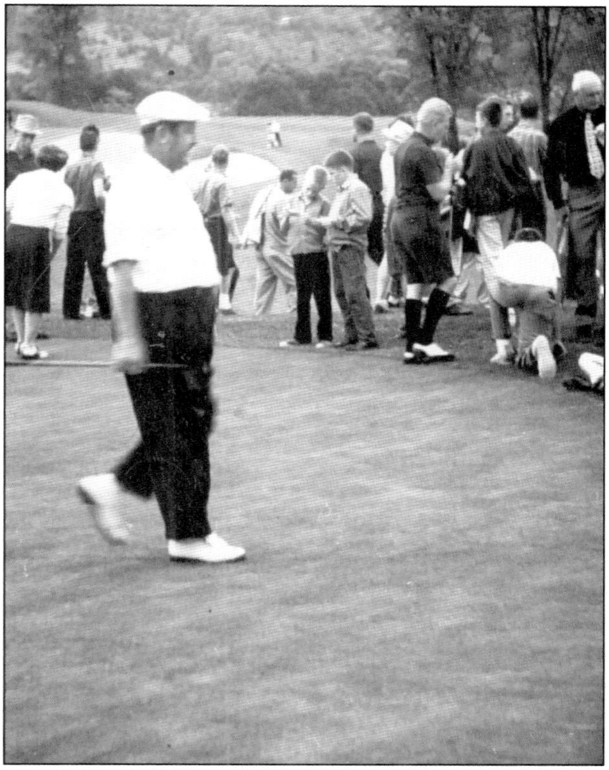

Talk of a golf course actually began in 1901, but it would be 1952 before the Crown Point Country Club was incorporated. In 1958, the club voted to add another nine holes. On August 30, 1959, four-time British Open champion Bobby Locke (center, with the white cap) was invited to the club and gave a golfing demonstration at the dedication of the expanded course. (CPCC.)

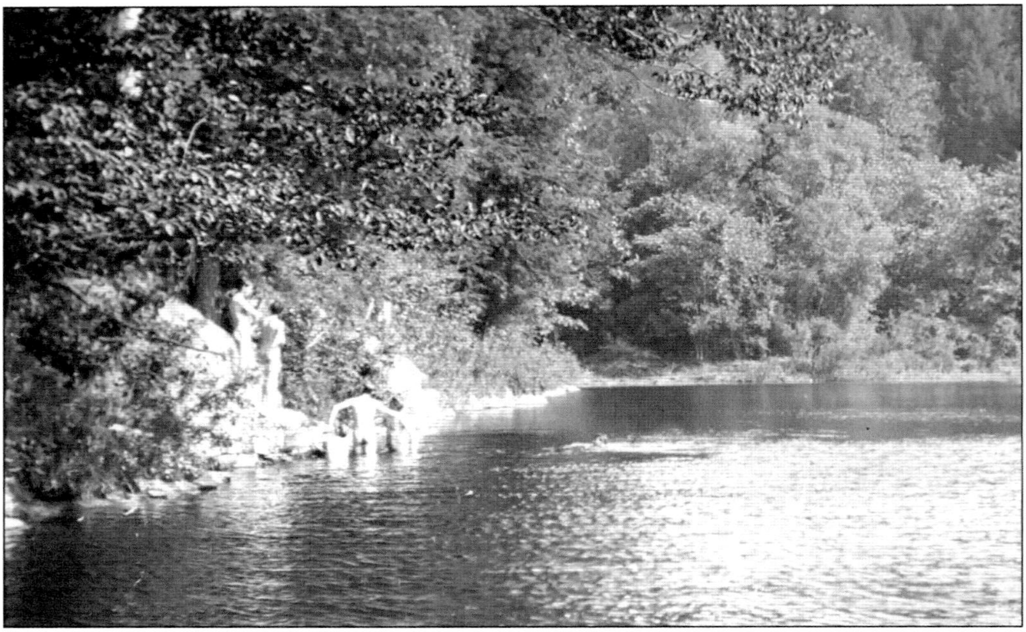

Sometimes, one had to find a place to cool off. Although it was not an official swimming area, many found the rocks in the Oxbow at the end of Pearl Street a suitable place for a dip. These boys apparently thought they had the spot all to themselves and did not expect any young ladies to show up. (WGS.)

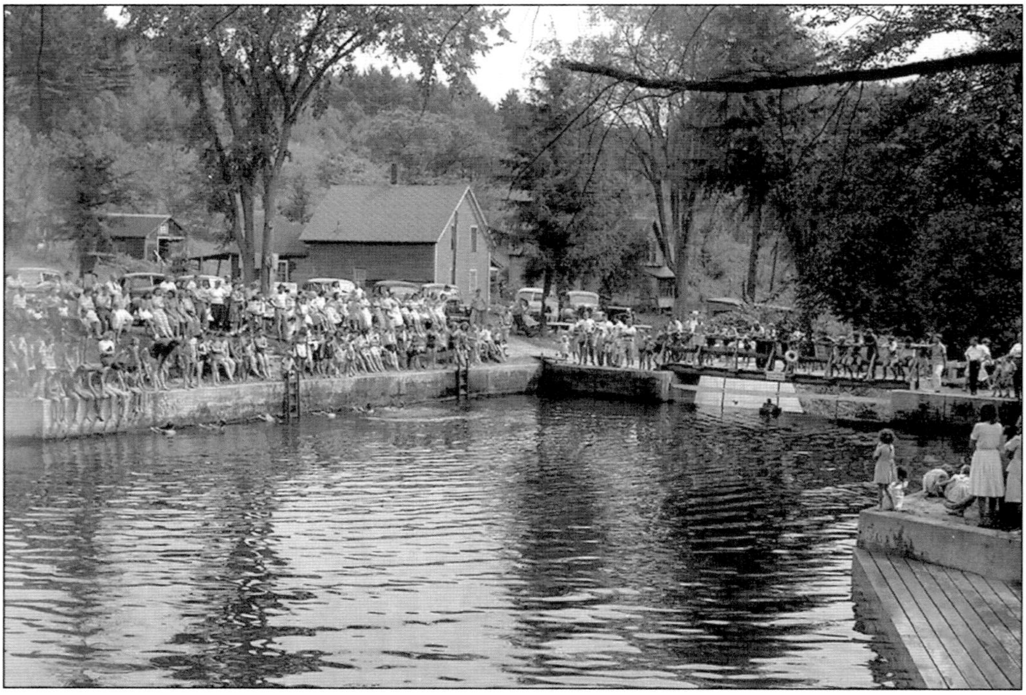

The town swimming area had been in the river near the baseball diamond at Riverside Park. In the late 1930s, it was decided that a real swimming pool was needed. The brook was dammed, the water treated, the Lions Club donated a slide, and in July 1938, this pool on Valley Street was opened. It remained the town pool until 1954 when the pool at Riverside was opened. (BL.)

*Eight*

# Charles Lindbergh's Visit

Once again, James Hartness was at the forefront of a new idea. He was convinced that aviation was the future, and by 1916, he had the first pilot's license in Springfield—and probably in Vermont. He was so convinced that the airplane would revolutionize transportation that he personally purchased property in North Springfield to establish an airport. It was going to cost money to prepare the field, and few people could readily grasp his vision. In spite of that, he asked the community, and it gave. After all, he had built the machine tool industry—and they were not going to let him down.

Although Hartness never owned an airplane, he was a national figure in the promotion of aviation, from consulting in Washington, DC, to locating and overseeing the construction of the airfield at Fort Ethan Allen in Colchester, Vermont, when he was governor. This would be important because, in 1927, after Lindbergh made his historic transatlantic flight, Springfield would be the only Vermont stop on his countrywide tour to promote aviation.

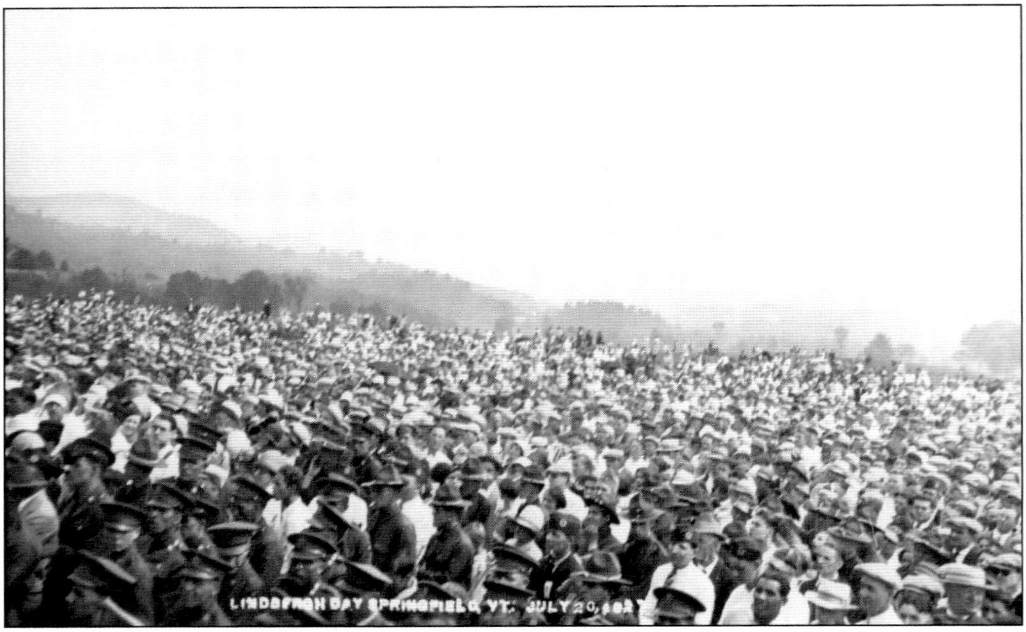

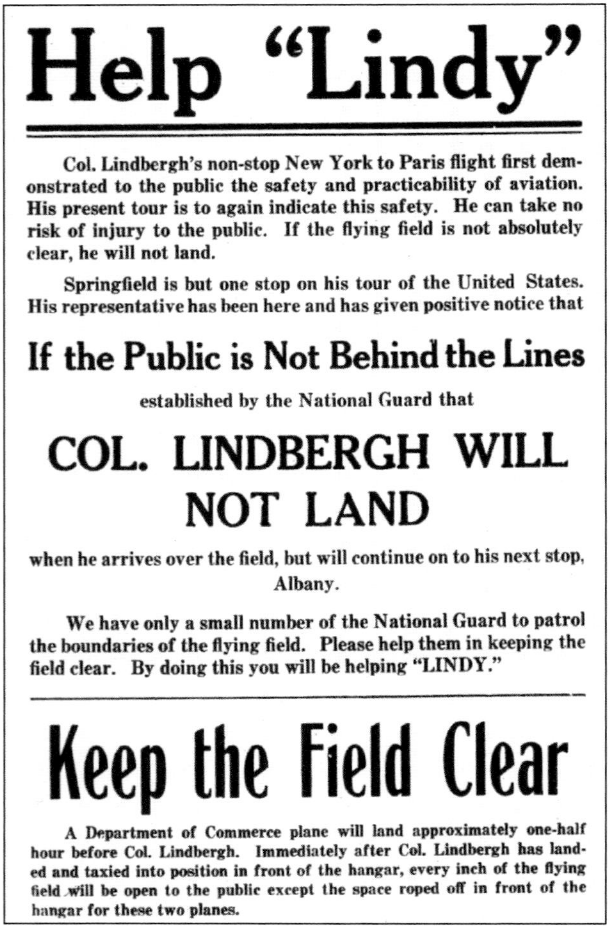

# Help "Lindy"

**Col. Lindbergh's non-stop New York to Paris flight first demonstrated to the public the safety and practicability of aviation. His present tour is to again indicate this safety. He can take no risk of injury to the public. If the flying field is not absolutely clear, he will not land.**

**Springfield is but one stop on his tour of the United States. His representative has been here and has given positive notice that**

## If the Public is Not Behind the Lines

**established by the National Guard that**

## COL. LINDBERGH WILL NOT LAND

**when he arrives over the field, but will continue on to his next stop, Albany.**

**We have only a small number of the National Guard to patrol the boundaries of the flying field. Please help them in keeping the field clear. By doing this you will be helping "LINDY."**

## Keep the Field Clear

**A Department of Commerce plane will land approximately one-half hour before Col. Lindbergh. Immediately after Col. Lindbergh has landed and taxied into position in front of the hangar, every inch of the flying field will be open to the public except the space roped off in front of the hangar for these two planes.**

On July 26, 1927, with only two weeks' notice, the town handled an estimated crowd of 30,000 people who came to welcome Charles A. Lindbergh. Roads were designated as one way, parking was created for automobiles, and first aid and comfort facilities were set up. Food was provided for the hundreds of volunteers who managed the crowd and kept people off the airfield. (HSP.)

As can be seen by this poster, managing the crowds was a very important job. There were very strict guidelines, and no one wanted to take the chance that those rules would be violated and Lindbergh would just fly over and keep on going. National Guard, local and area police, and hundreds of volunteers made it their business to see to it that the rules were followed. (HSP.)

The crowds got excited when a plane was spotted coming up the valley, but it turned out to be Lindbergh's forward team. Finally, the *Spirit of St. Louis* appeared, circled the field several times, and landed at about 2:00 p.m. He looked tired and did not smile until Governor Hartness spoke to him, and then the famous "Lindy" smile appeared. (HSP.)

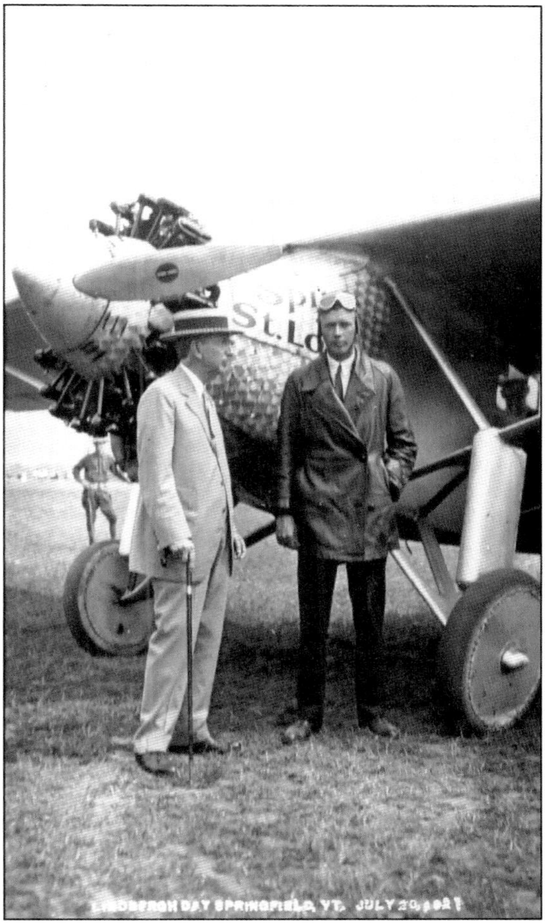

At the field, a speaker's stand with amplifiers was set up by the hanger. During his address to the crowd, Lindbergh said that Vermont's hilly terrain made flying a better means of transportation because it was easier and faster to fly over the hills than to drive around them. (HSP.)

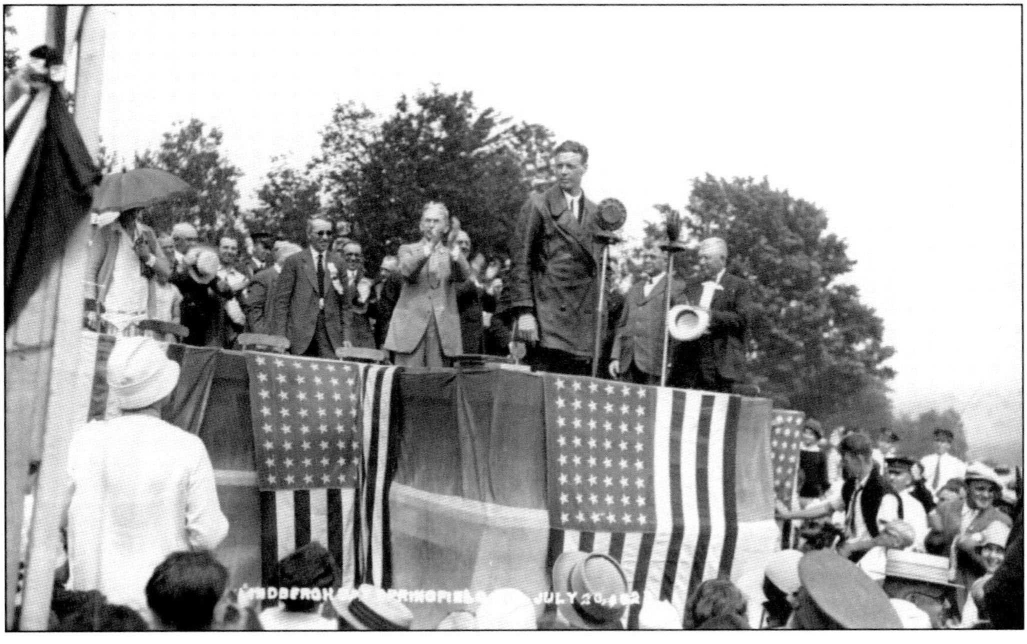

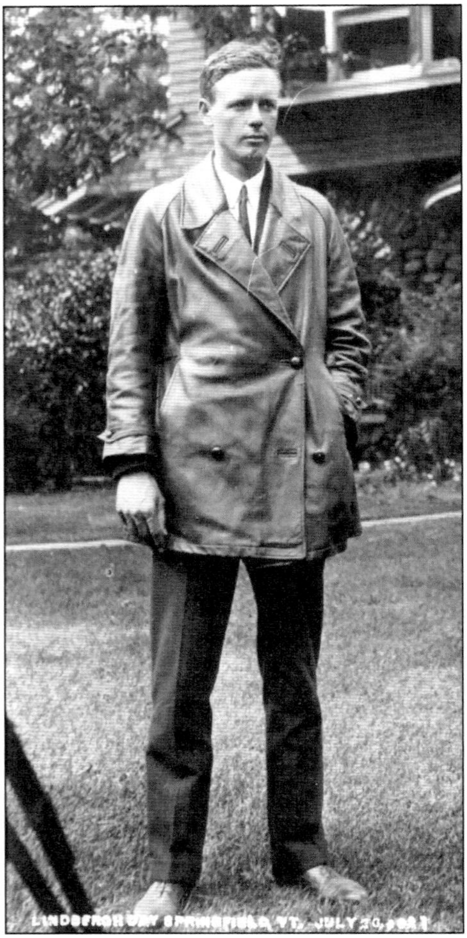

A banquet was held at the Community House for 500 selected guests interested in promoting commercial aviation, the theme of the tour. Amplifiers were set up in the windows of the Community House so that people gathered outside could hear the speeches. Lindbergh spent the night at the Hartness residence—believed to be the only private home he stayed in on the trip. (HSP.)

The next morning, Lindbergh took off at 10:20 a.m. Flying over a crowd of about 2,000, he performed some hair-raising banks and climbs and then straightened his course for Plymouth to pay tribute to Vermont's favorite son, Pres. Calvin Coolidge. He then headed off to Albany, New York. (HSP.)

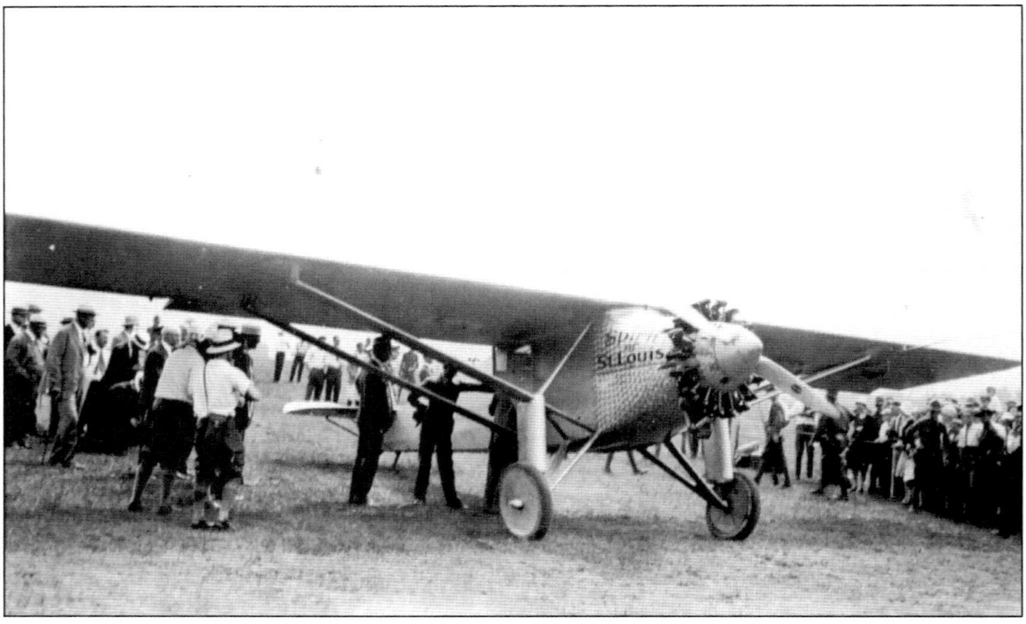

# *Nine*

# STELLAFANE

Russell Porter was born in Springfield in 1871. An artist, explorer, and telescope maker, his interest in the sky began on arctic and subarctic expeditions between 1894 and 1906. Finally settling in Port Clyde, Maine, he had continuing correspondence with James Hartness, who supported his interest in telescopes both intellectually and financially. In 1913, Hartness had a set of 16-inch discs sent to him to make into mirrors.

In early 1919, Hartness asked Porter to come to Springfield to help him with an optical invention he was working on—the optical comparator. This was Porter's chance to work on his hobby and get paid to do it. He accepted and moved back to his hometown.

In addition to helping Hartness perfect the comparator, on August 17, 1920, Porter brought together the first group of amateur telescope makers, 16 men and one woman who wanted to make their own telescopes. Hartness provided space at the Main Street J&L building, and Porter instructed the group in mirror making. First, they had to learn to grind it to the correct depth. Then, many more hours of grinding with finer abrasives were required. Finally, the mirrors were polished with a tool of pitch covered with a slurry of very fine powder called rouge. This was the beginning of Springfield amateur telescope makers and their clubhouse known as Stellafane.

James Hartness would have liked to establish the Porter Comparator Company, but Russell Porter was not interested in running a business. Instead, in 1928, he received a telegram from the California Institute of Technology inviting him to Pasadena to work on the 200-inch Palomar Telescope. He would spend the remainder of his life in California.

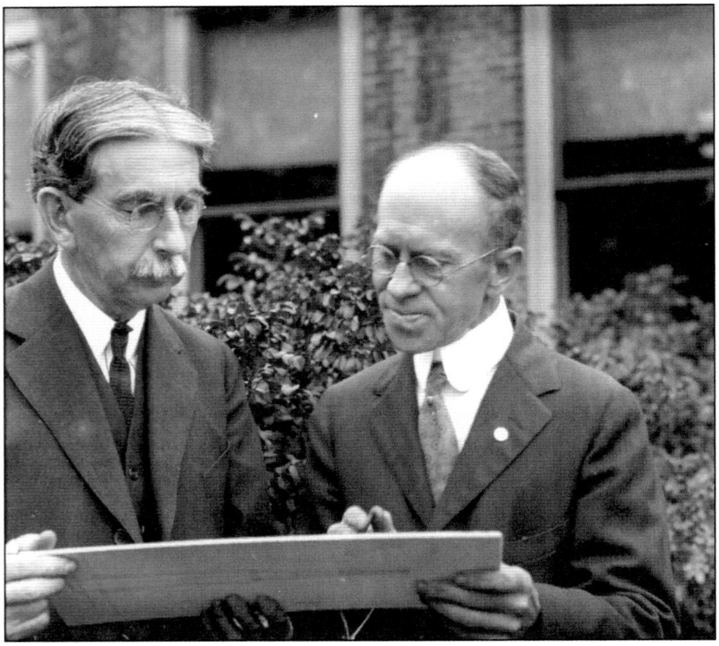

In this photograph are good friends James Hartness (left) and Russell Porter. Although Porter left Springfield, his heart was always there with his amateur telescope makers. He would come back as often as possible over the years and especially enjoyed attending the annual Stellafane Conventions. (Stellafane Telescope Makers.)

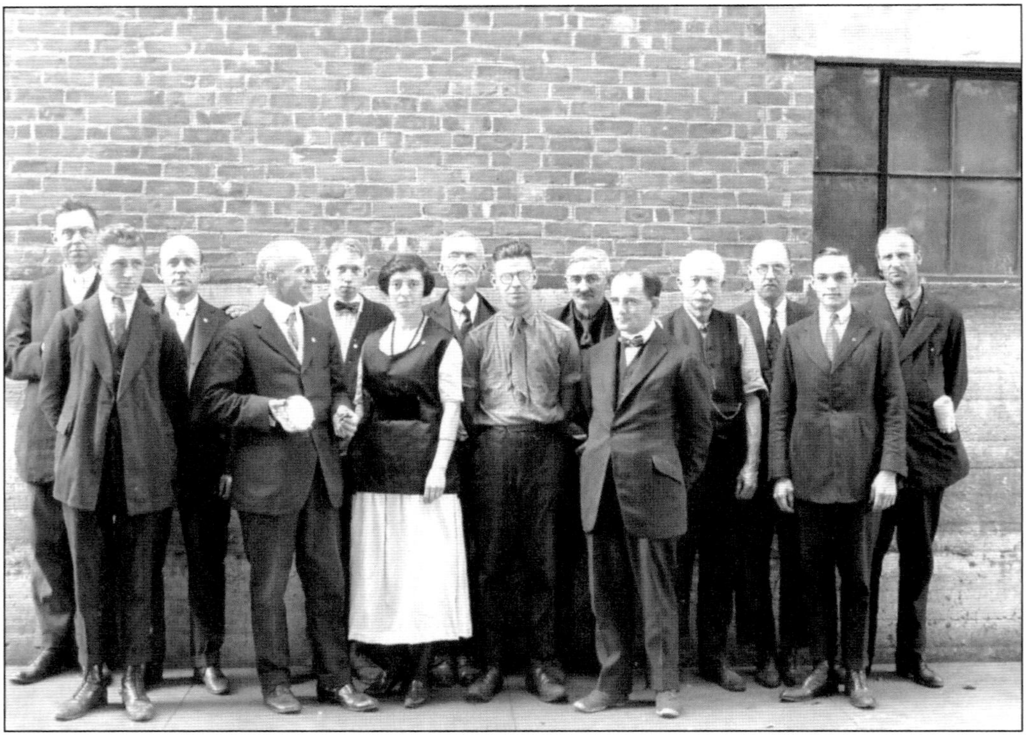

The first class of telescope makers included in this photograph are, from left to right: Oscar Marshall, Clyde Baldwin, Eugene Lockwood, Russell Porter, Raymond Fairbanks, Gladys Piper, Oscar Fullam, Ralph Baker, Frank Whitney, Roy Lyon, Everett Redfield, Carlton Damon, Ernest Brookings, and Guy Baker. Not included in this photograph are Albert Herrick, Charles Longe, and John Pierce. Although Gladys moved away in 1921, leaving it an entirely male group, she did complete her telescope. (STL.)

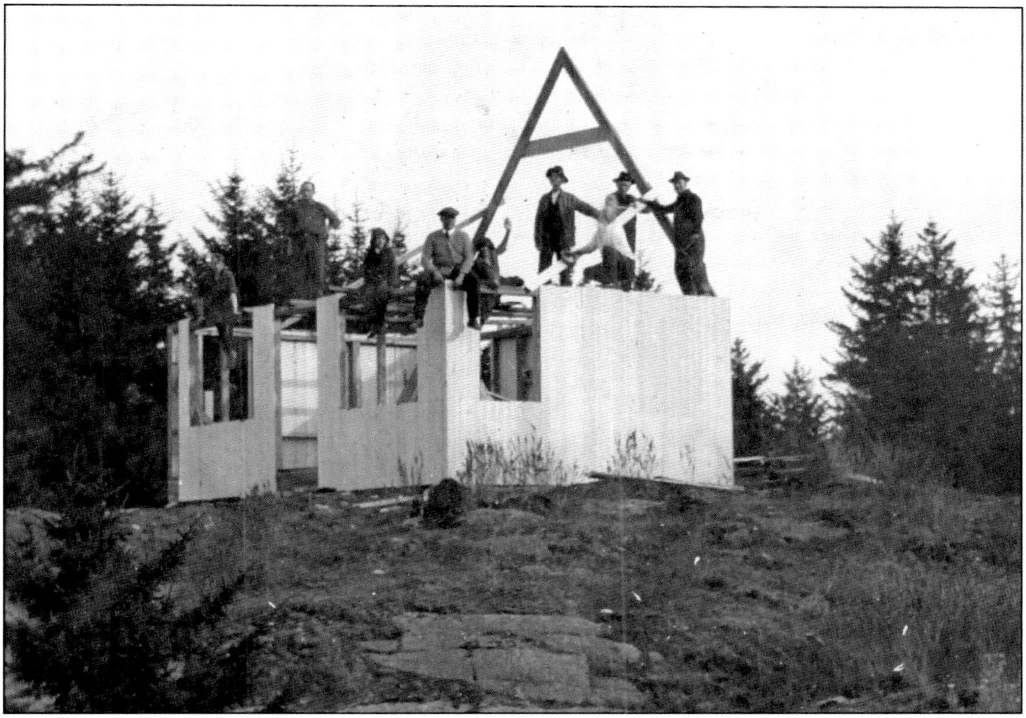

The first official meeting of the Springfield Telescope Makers club was on December 7, 1923. It was determined that one of the requirements for membership would be to successfully make a speculum (mirror). A more important discussion at that meeting was the completion of the bungalow that had been started in the fall. People were tired of lugging camping gear and telescopes to the site. (STL.)

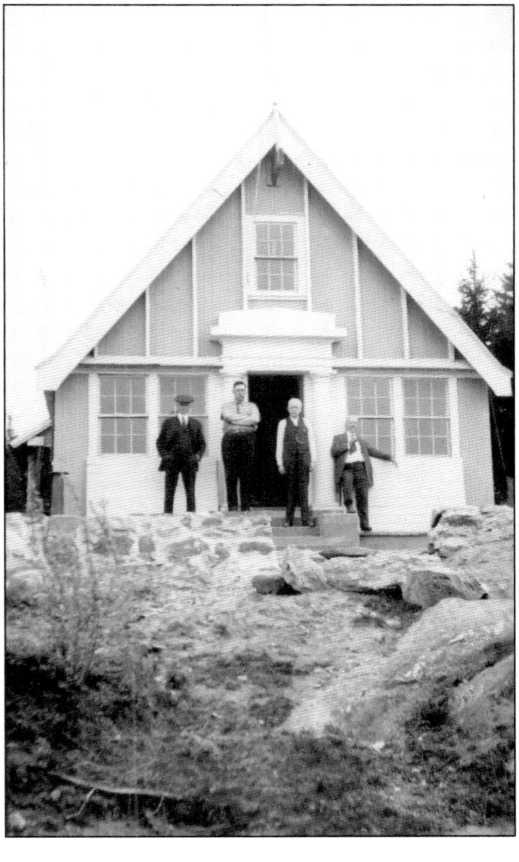

Porter had inherited the land on top of Breezy Hill. He offered to lease the land to the club and finally sold it to them for $1 in 1928. Always looking to save money, the club agreed to accept free paint for the building. It turned out to be pink. The clubhouse continues to be painted pink to this day. Included in this photograph are, from left to right, unidentified, Oscar Marshall, Everett Redfield, and Russell Porter. (STL.)

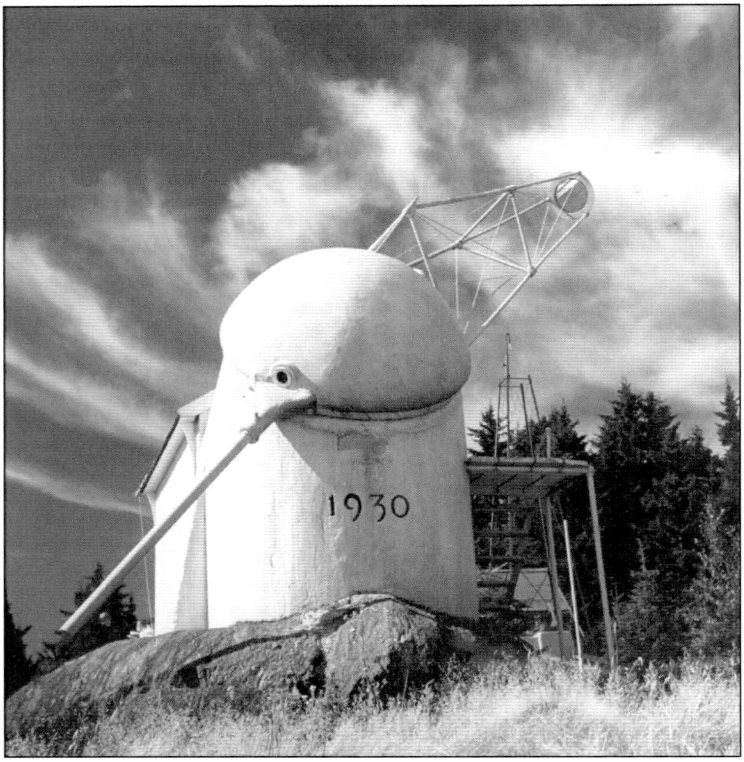


Porter left Springfield in 1928 but kept in touch and attended conventions. In September 1929, he brought plans for a telescope to be erected in front of the clubhouse. A motorized turret telescope modeled after Hartness's, it would use Porter's 16-inch mirrors from Port Clyde and be the first reflecting turret telescope in the world. By the 1930 convention, the telescope was in place but without the motor drive. (STL.)

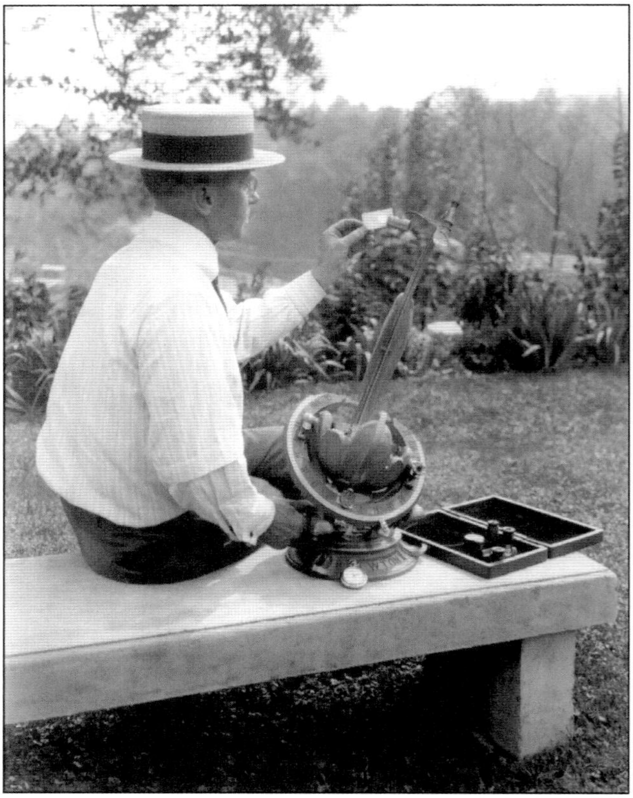

The Garden Telescope was designed to be a permanent, multipurpose telescope and sundial that could be used to observe the stars and the nearby landscape, as well as being ornamental. In this photograph, Porter is demonstrating it as a sundial. Notice the pocket watch on the bench. His hope was that it could be built so that it would be affordable for people of modest means. (Stellafane Telescope Makers.)

*Ten*

# THE 1927 FLOOD AND THE 1938 HURRICANE

The rains that caused the 1927 flood started on Wednesday, November 2, and by Thursday evening, people on Clinton Street began calling for help. The electricity went off and then the gas, and early Saturday the water main north of town broke. The Black River went over its banks, and the water was four feet deep in the Clinton Street area where the most damage was done. The electric substation was damaged, the bridge on Bridge Street went out, and Fellows Gear Shaper was left with a muddy mess on their main floor. The water also backed up two feet deep in the lower part of J&L. The railroad had two washouts, and the B&M railroad to Charlestown was out. The telephone company maintained service using the generator of an automobile parked outside the Bank Block. No lives were lost, but the total damage was estimated at $250,000.

The last two pictures are of the 1938 hurricane, the first ever recorded in Springfield. It started to rain on Sunday, and by Wednesday, September 21, the heavy rain and winds reached hurricane velocity. Trees were uprooted all over town, damaging about 60 percent of the telephone lines. As in the flood of 1927, a water main broke, adding to the problems. Clinton Street, one of the low spots in town, was again flooded. The water crested five feet below the level reached during the flood of 1927.

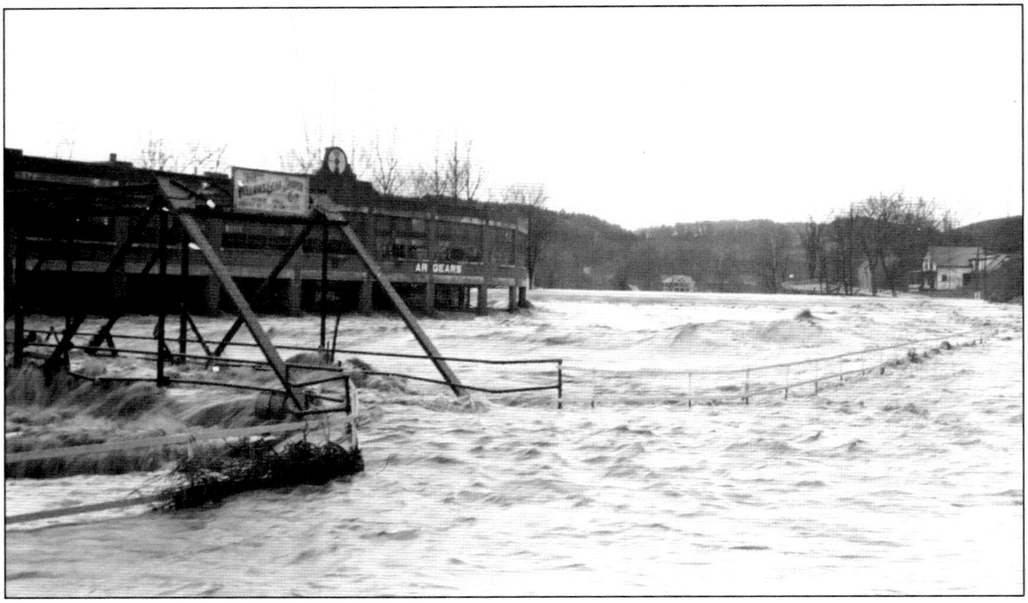


In this image taken during the flood, River Street truly was a river. One can barely make out the dam to the right of Fellows. Water ran through the main floor of the building. There were major washouts all along the road. Notice that the sign over the bridge reads "cross at your own risk." (STL.)

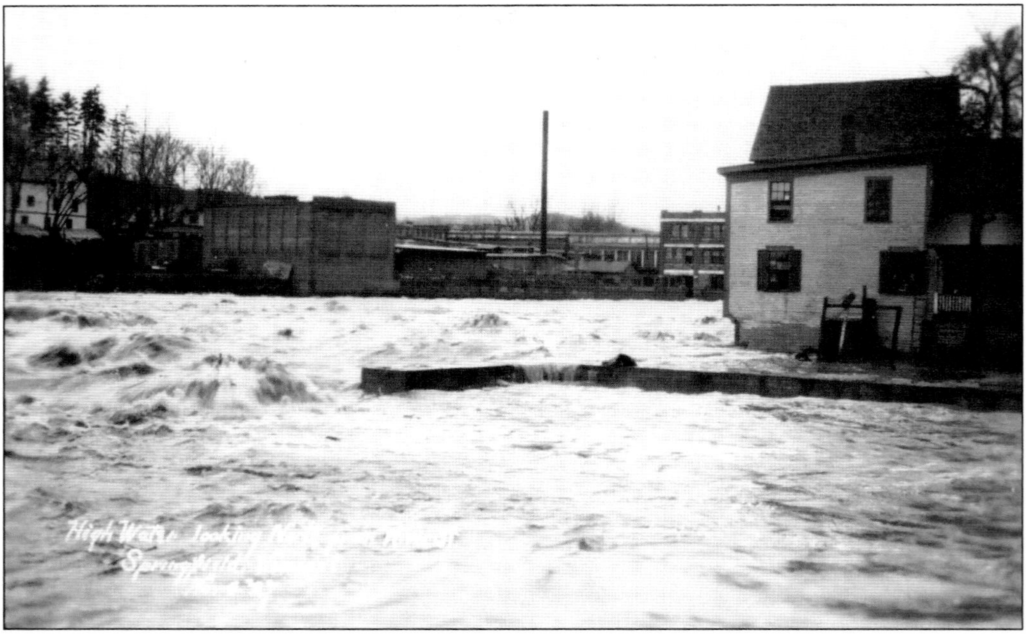

This photograph was taken near the foot of Elm Hill looking back towards Fellows Gear Shaper. The water was across the entire road, making it look more like a lake than a river. (STL.)

Hawkins Flower Shoppe was on the Park Street side of Falls Bridge. This photograph was taken before the current addition was put onto the front of that building. The structure's saving grace was that the water was running away from the door and not towards it. (STL.)

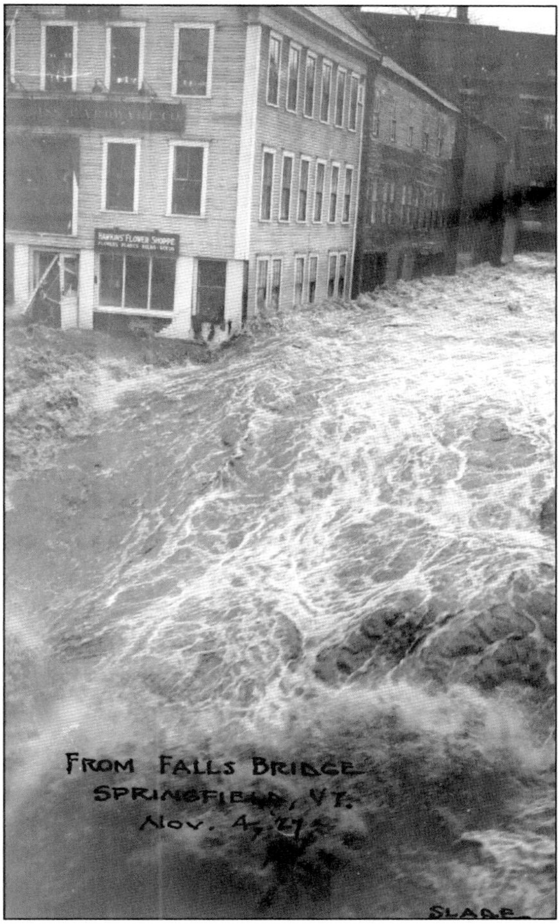


With water rushing over the falls and given the force of the floodwaters, it was amazing that the buildings along the river held up like they did. Notice the little shed on the right below the bridge. Even in good weather, that little building seemed to be clinging to the side of the rocks. (STL.)

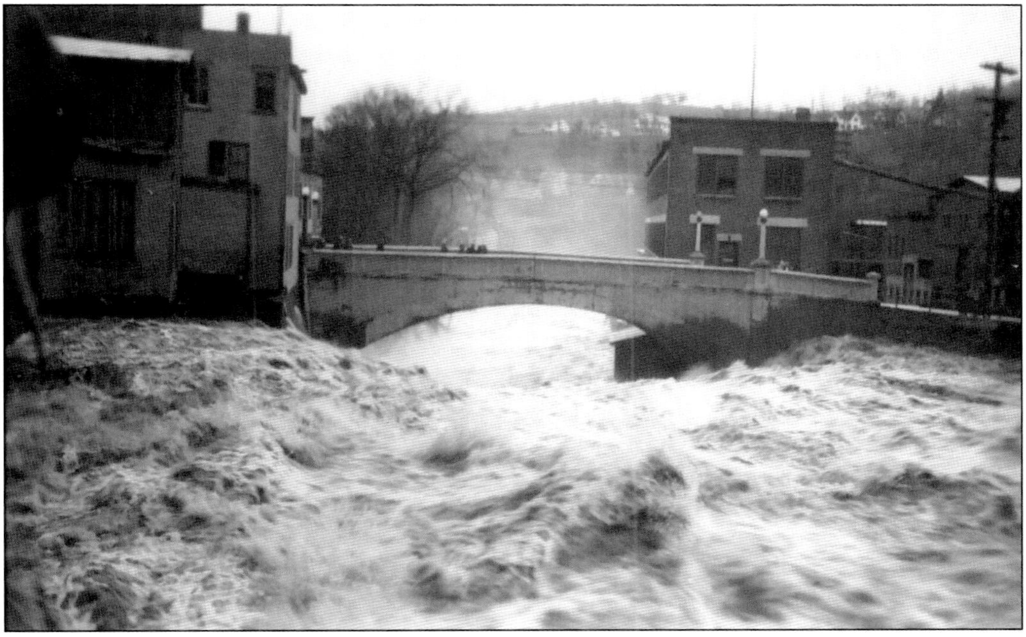

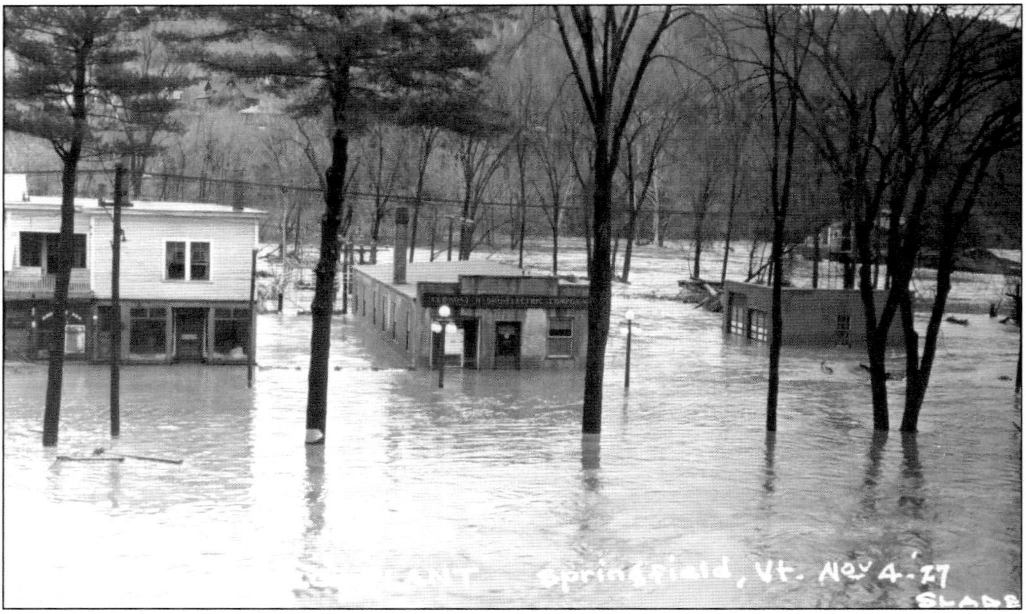


Clinton Street is underwater in this photograph. The building in the middle was the Springfield Hydro-Electric Corporation, and behind it was the power plant. (STL.)

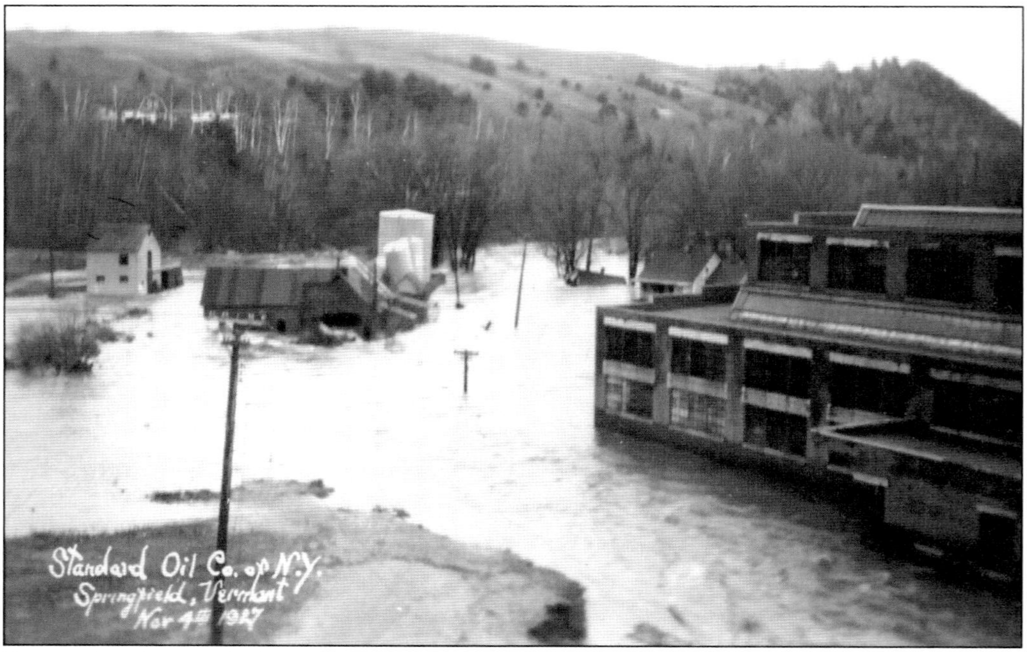


The road with the telephone poles at the bottom of the photograph is Hartness Avenue (by the Fire Station) heading towards Bridge Street. Standard Oil Company of New York is in the center. The foundry is on the right, and the river was normally on the other side of it. (STL.)

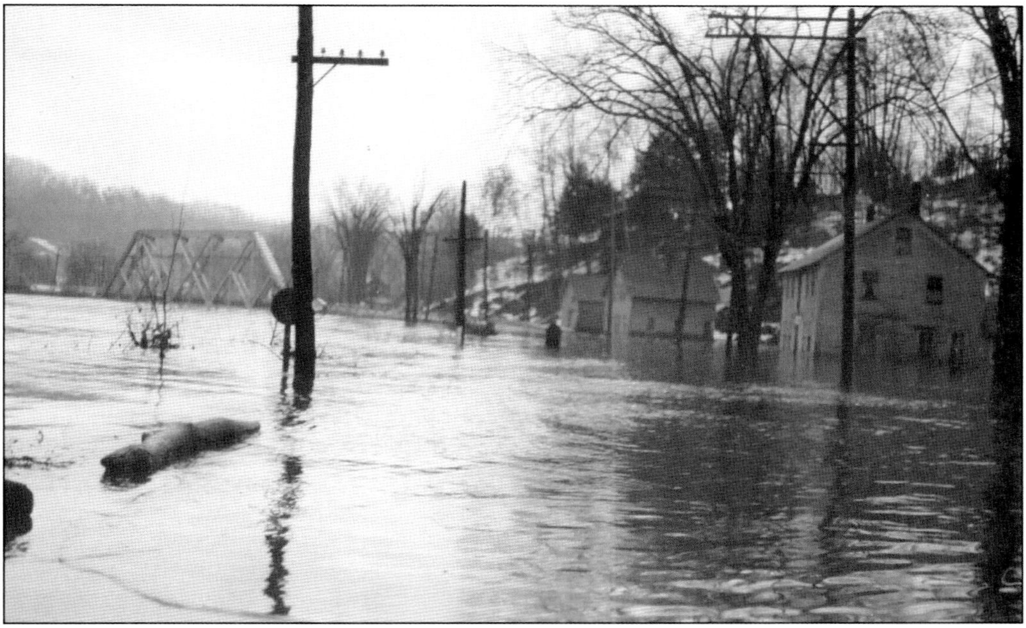

In this photograph, looking down river near the Toonerville Bike Trail (the old railroad bed), the Paddock Road Bridge can be seen in the distance. The old Goulds Mills power dam was on the left, and Randall Hill Road was out of the frame to the right. There appears to be a man standing in what should be the middle of the road. (STL.)

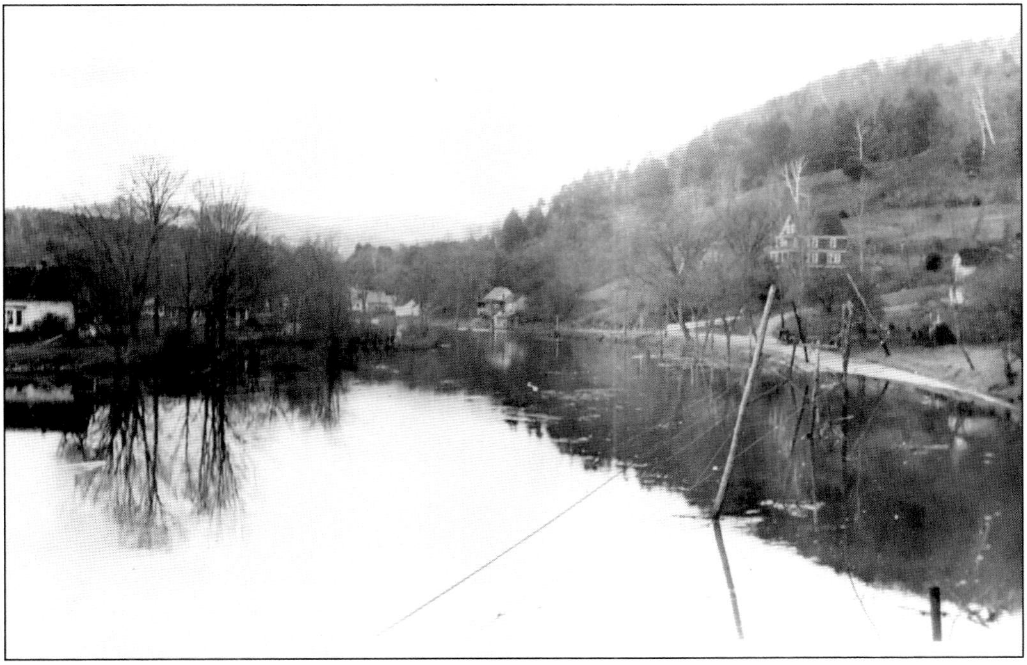

Paddock Road (the former main thoroughfare into town) is on the right in this photograph. The first street heading off to the right is Spencer Hollow Road. Perley Gordon Road is on the left side of the river and was a dead-end street. Not only were the road and railroad tracks underwater, the Goulds Mills Bridge had water up to its decking. The road above the bridge was also washed out. (STL.)

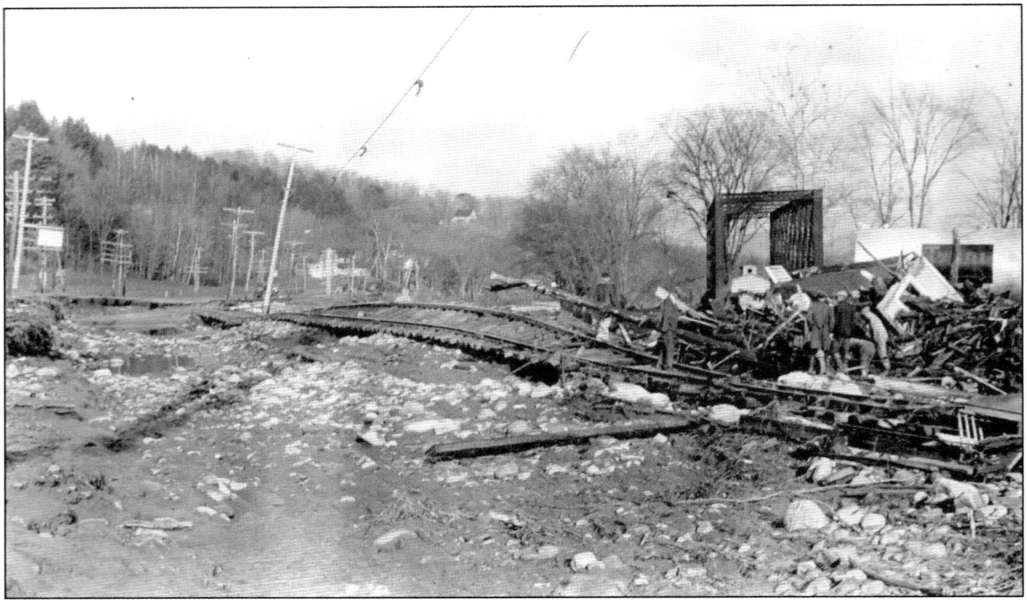

This photograph was taken after the flood. This bridge crossed the river just up from Seaver's Brook Road to the old brass foundry that Frank Corliss had torn down in the summer and replaced with tourist cabins. The bridge was never replaced. Today, the water treatment plant is on the left. (STL.)

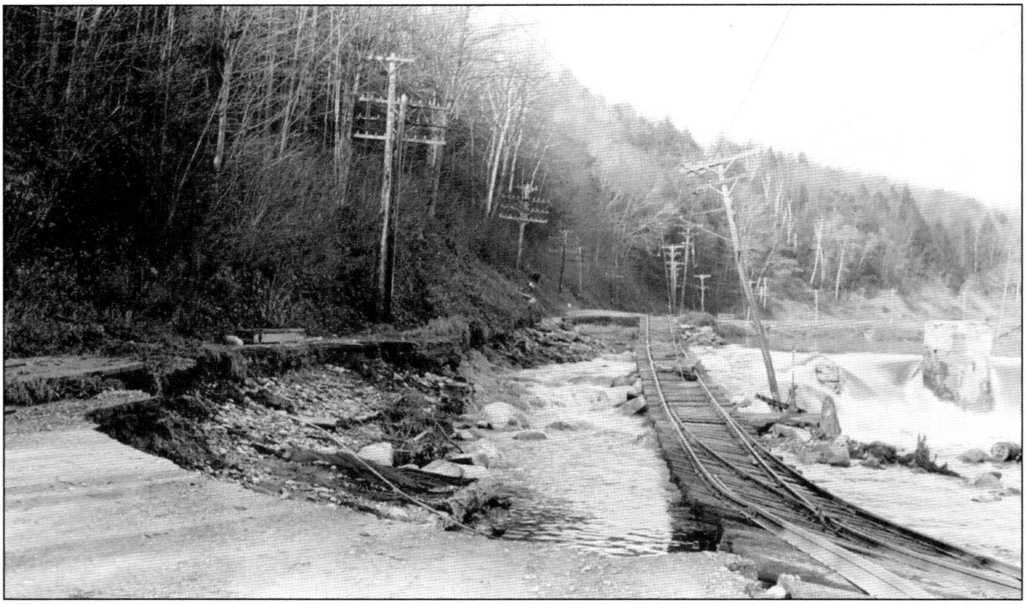

The track and road washout in Goulds Mills was captured in this post-flood image. The foot of Randall Hill is on the left, and the old power dam is on the right. This is where the Toonerville Walking Path is today. (STL.)

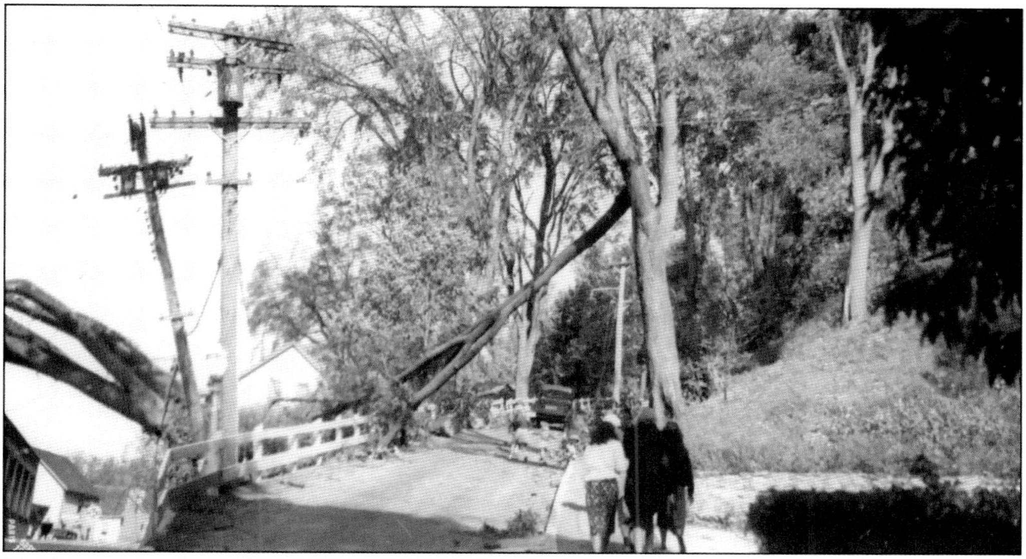

The hurricane of 1938 downed trees, which formed dangerous arches over the roads at the corners of Main and River Streets and on Elm Hill. (STL.)

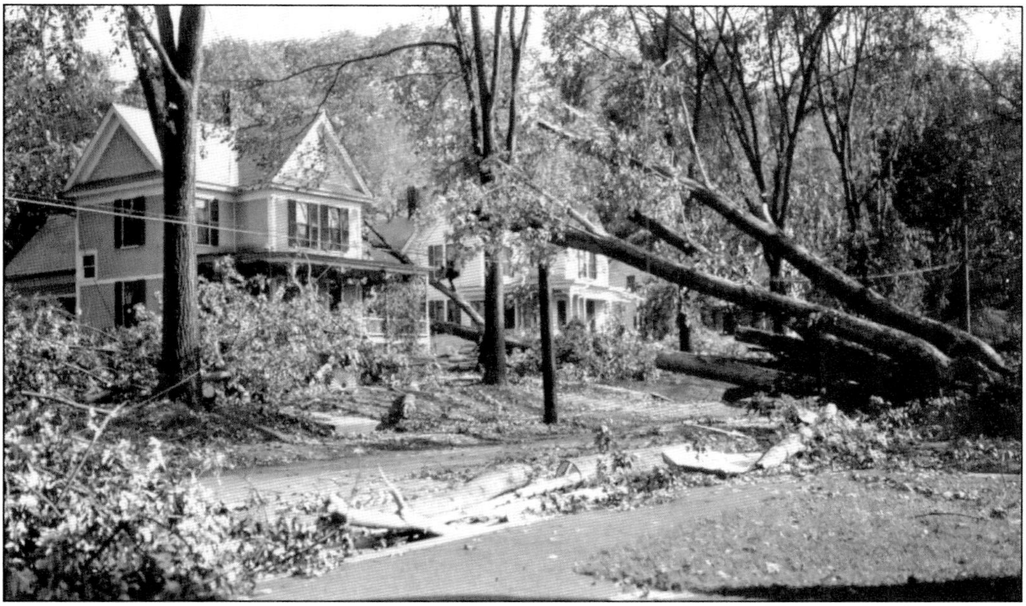

This was just some of the damage done on Summer Street by the hurricane. These were not the only downed trees on the street or in town. (STL.)

Consistent with our mission to preserve history on a local level, this book was printed in South Carolina on American-made paper and manufactured entirely in the United States. Products carrying the accredited Forest Stewardship Council (FSC) label are printed on 100 percent FSC-certified paper.